# DIGITAL ELECTRONICS:
## Theory and Experimentation

# DIGITAL ELECTRONICS:

## Theory and Experimentation

FREDRICK W. HUGHES
*Electronics Training Consultant*

PRENTICE-HALL, Englewood Cliffs, New Jersey 07632

*Library of Congress Cataloging-in-Publication Data*

Hughes, Fredrick W., 1936–
    Digital electronics.

    1. Digital electronics. 2. Digital electronics—
Laboratory manuals. I. Title.
TK7868.D5H825 1986      621.3815      85-19186
ISBN  0-13-212556-0

Editorial/production supervision
    and interior design: Tom Aloisi
Cover design: George Cornell
Manufacturing buyer: Gordon Osbourne

*This manual is dedicated to all technical authors
that have given the time and effort required to
researching, testing and reviewing materials for
such a book.*

Printed in the United States of America

10  9  8  7  6  5  4  3

ISBN: 0-13-212556-0  025

Prentice-Hall International (UK) Limited, *London*
Prentice-Hall of Australia Pty. Limited, *Sydney*
Editora Prentice-Hall do Brasil, Ltda., *Rio de Janeiro*
Prentice-Hall Canada Inc., *Toronto*
Prentice-Hall Hispanoamericana, S.A., *Mexico*
Prentice-Hall of India Private Limited, *New Delhi*
Prentice-Hall of Japan, Inc., *Tokyo*
Prentice-Hall of Southeast Asia Pte. Ltd., *Singapore*
Whitehall Books Limited, *Wellington, New Zealand*

# Contents

## 4 CLOCK AND TRIGGER CIRCUITS                                                          97

## 5 FLIP-FLOPS                                                                            133

## 6 BINARY REGISTERS                                                                      156

## 7   BINARY COUNTERS                                                                        184

## 8   ARITHMETIC CIRCUITS                                                                    208

## 12 DIGITAL IC OPERATION AND SPECIFICATIONS                                307

## APPENDIX: BASIC DIGITAL TEST EQUIPMENT                                335

# Preface

This manual can be used with existing textbooks and digital training equipment. It can be used as a practical laboratory workbook for training in the use and understanding of digital circuits. The manual can also be used as a stand-alone text and/or laboratory manual or workbook for teaching the basics of digital electronics.

A step-by-step method is used throughout the book to enable the reader to master each section so that the accumulation of knowledge can be used in each succeeding section in a building-block fashion, from basic to more complex. The material is presented in a repetitious manner, which is essential in acquiring a skill, but with varying methods which are easy, challenging, and fun to perform.

The book is designed as an individualized learning package and involves the reader in the activities of learning. Many illustrations are included to familiarize students with logic gate symbols and the operation of various devices and circuits. It provides the instructor with student-centered instructional material which does not require preparation.

Each unit follows the same format so the reader can become accustomed to the learning procedure. The first section introduces the topic and provides basic theory for a fundamental understanding. Section 2 consists of basic drawing exercises for familiarity of symbols, concepts, and operating procedures. In Section 3 a definition exercise provides a review of the topics covered. Section 4 involves experiments to enable students to gain manipulative skills and knowledge in working with digital circuits. Fill-in questions are given at the end of each experiment to emphasize the important points gained from the experiment.

An instant review of the topics covered in the first part of the unit is given in Section 5 to reinforce the student's essential knowledge. Section 6 involves more exercises on basic principles covered and some contain more complex circuits. A basic troubleshooting application on digital devices is provided in Section 7 to further the student's hands-on skills. Section 8 has two self-checking quizzes to aid students in gaining experience taking digital electronic tests and as a final review. Answers to the experiments and quizzes follow each unit.

## A SPECIAL MESSAGE TO STUDENTS

You are studying digital electronics to obtain employment or upgrade yourself in a lucrative, exciting, and challenging industry. It is up to you, not your instructor (or anyone else, for that matter), how well you master the knowledge and skills required for such a fascinating mainstream career. To gain full value from this manual:

1. Perform all the experiments, exercises, and self-checking quizzes.

2. Review each section of the book from time to time to refresh your memory as to the basic concepts of the various logic gates, flip-flops, and digital devices.

3. Keep this book and review it before you take a job entrance examination.

4. Refer to this book when you are on the job to become a more competent technician and to prepare for a higher-level position.

5. Have this book handy when you are servicing equipment, to refer to the sections on testing digital logic devices and troubleshooting circuits. It can serve as a valuable aid to reducing time spent on repairing electronic devices.

6. Review the material in this book when you are preparing to take professional license examinations, such as those for the FCC radio telephone operator's license and the CET (Certified Electronic Technician).

## PERFORMING EXPERIMENTS

Most, if not all experiments can be performed on the commercial logic trainer boards existing in most schools. If these trainers are not available, the experiments can still be performed by constructing the inexpensive digital test equipment described in the appendix and using breadboards with 1-mil-on-center holes.

## SPECIAL NOTE TO INSTRUCTORS

Some of the more complex circuits in the experiments can be hard-wired on perforated board modules (it is advisable to use IC sockets). To meet laboratory time requirements, these modules can be passed out to groups of students to use in performing the experiments.

*Fredrick W. Hughes*

# Unit A
# Preliminary Review

---

# Safety

## SECTION A-1
## INTRODUCTION

The experiments in this manual do not use a voltage regulator greater than + 15 V dc; therefore, the chance of getting an electrical shock is greatly reduced. However, all voltages do have the potential to burn materials and start fires, to destroy electronic components, and present hazards to the person performing the operations. Common sense and an awareness of electrical circuits is important whenever you are working on these experiments. An electronic technician or student may have to work with high voltages, power tools, and machinery. Before actual work is performed, sufficient instruction should be acquired in the proper use and safety requirements of all electronic devices.

## SECTION A-2
## CURRENT HAZARDS AND VOLTAGE SAFETY PRECAUTIONS

It takes a very small amount of current to pass through the human body from an electrical shock to injure a person severely or fatally. The 60-Hz current values affecting the human body are as follows:

| Current Value | Effects |
| --- | --- |
| 1 mA (0.001 A) | Tingling or mild sensation |
| 10 mA (0.01 A) | A shock of sufficient intensity to cause involuntary control of muscles, so that a person cannot let go of an electrical conductor |
| 100 mA (0.1A) | A shock of this type lasting for 1 second is sufficient to cause a crippling effect or even death |
| Over 100 mA | An extremely severe shock that may cause ventricular fibrillation, where a change in the rhythm of the heartbeat causes death almost instantaneously |

The resistance of the human body varies from about 500,000 Ω when dry to about 300 Ω when wet (including the effects of perspiration). In this case, voltages as low as 30 V can cause sufficient current to be fatal ($I =$ voltage/wet resistance $= 30$ V/$300$ Ω $= 100$ mA).

Even though the actual voltage of a circuit being worked on is low enough not to present a very hazardous situation, the equipment being used to power and test the circuit (i.e., power supply, signal generator, meters, oscilloscopes) is usually operated on 120 V ac. This equipment should have (three-wire) polarized line cords that are not cracked or brittle. An even better safety precaution is to have the equipment operate from an isolation transformer, which is usually connected to a workbench. To minimize the chance of getting shocked, a person should use only one hand while making voltage measurements, keeping the other hand at the side of the body, in the lap, or behind the body. Do not defeat the safety feature (fuse, circuit breaker, interlock switch) of any electrical device by shorting across it or by using a higher amperage rating than that specified by the manufacturer. These safety devices are intended to protect both the user and the equipment.

## SECTION A-3
## NEAT WORKING AREA

A neat working area requires a careful and deliberate approach when setting it up. Test equipment and tools should be set out on the workbench in a neat and orderly manner. Connecting wires from the test equipment to the circuit under test should be placed so as not to interfere with testing procedures.

Before power is applied to a circuit, the area around the circuit should be cleared of extra wires, components, hand tools, and debris (cut wire and insulation).

## SECTION A-4
## HAND TOOL SAFETY PRECAUTIONS

Hand tools can be dangerous and cause severe injuries. Diagonal cutters, wire strippers, long-nose pliers, and crimping tools can pinch and cut. Use care in cutting wire since small pieces can become projectiles and hit another person in the face or eye.

Screwdrivers should be held properly so that they do not slip and puncture some part of the body. Do not use them as chisels or cutters.

A soldering iron should have a holder on which to place it. Care must be used not to burn the body or other materials. Be careful of hot solder, which can splash and cause severe burns, especially to the eyes and face.

## SECTION A-5
## IN CASE OF ELECTRICAL SHOCK

When a person comes in contact with an electrical circuit of sufficient voltage to cause shock, certain steps should be taken as outlined in the following procedure:

1.  Quickly remove the victim from the source of electricity by means of a

switch, circuit breaker, pulling the cord, or cutting the wires with a well-insulated tool.

2. It may be faster to separate the victim from the electrical circuit by using a dry stick, rope, leather belt, coat, blanket, or any other nonconducting material.

   *CAUTION:* Do not touch the victim or the electrical circuit unless the power is off.

3. Call for assistance, since other persons may be more knowledgeable in treating the victim or can call for professional medical help while first aid is being given.

4. Check the victim's breathing and heartbeat.

5. If breathing has stopped but the victim's pulse is detectable, give mouth-to-mouth resuscitation until medical help arrives.

6. If the heartbeat has stopped, use cardiopulmonary resuscitation, *but only if you are trained in the proper technique.*

7. If both breathing and heartbeat have stopped, alternate between mouth-to-mouth resuscitation and cardiopulmonary resuscitation (*but only if you are trained*).

8. Use blankets or coats to keep the victim warm and raise the legs slightly above head level to help prevent shock.

9. If the victim has burns, cover your mouth and nostrils with gauze or a clean handkerchief to avoid breathing germs on the victim and then wrap the burned areas of the victim firmly with sterile gauze or a clean cloth.

10. *In any case, do not just stand there*—do something within your ability to give the victim some first aid.

# Unit B
# Preliminary Review

# Basic Digital Logic
# Test Instruments

## SECTION B-1
## THE LOGIC PROBE

Testing analog (or linear) circuits usually involves a voltmeter to verify specific voltage levels at various points in a circuit. Very often, testing digital circuits requires only verifying the presence or absence of a voltage, not its specific level. Many digital circuits operate on a standard voltage level, such as +5, +12, or −12 V. Digital circuits are tested for 1s and 0s. In most circuits a 1 represents a positive voltage and a 0 represents zero voltage or ground.

The logic probe is a rather simple test instrument that detects the presence or absence of a voltage level. It is a small hand-held device as shown in Figure P-1. For proper operation the logic probe requires power, which it receives from the circuit being tested, by connecting the power leads to the appropriate power supply connections (assume that $+V_{CC} = +5$ V).

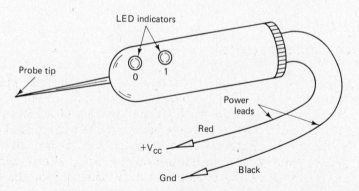

**Figure P-1** Logic probe.

The logic probe is used by placing the probe tip at the point in the circuit to be tested. If a positive voltage (must be of sufficient voltage, approximately 2 V) is present, the LED indicator labeled 1 will light. If the voltage is 0 V, the LED indicator labeled 0 will light.

Some logic probes may have only one LED indicator, in which case the presence of a positive voltage will cause it to light and 0 V will cause it to be out. Other logic probes may be more sophisticated and have LED indicators to detect pulses and for special purposes. The appendix shows a schematic diagram and procedure for constructing a basic TTL/CMOS logic probe.

## SECTION B-2
## THE LOGIC PULSER

The logic pulser shown in Figure P-2 is similar to the lobic probe and can be used as its counterpart. While the logic probe is used to test the presence of a 1 or 0, the logic pulser is used to inject a 1 or 0 into a digital circuit. A momentary-action pushbutton switch is used to activate a positive voltage at the probe tip. Like the logic probe, the logic pulser also receives its operating power from the circuit being tested by connecting the power leads to the appropriate power supply points.

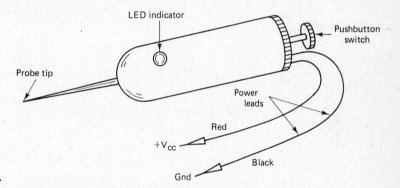

**Figure P-2**  Logic pulser.

The logic pulser is used by placing the probe tip at the desired input point in the circuit and then injecting a 1 by pressing the pushbutton switch. The LED indicator shows when the probe tip is at a 1 and 0 level; 1 = on, 0 = off. The logic pulser may also be called a digital signal injector. The appendix shows a schematic diagram and procedure for constructing a basic TTL/CMOS logic pulser.

## SECTION B-3
## USE OF THE LOGIC PULSER AND LOGIC PROBE

Figure P-3 shows a basic procedure for using the logic pulser and logic probe together to test a digital circuit. First, the power leads of each test device are connected to the circuit. Next, the logic pulser is placed at the input to the circuit and the logic probe is placed at the output of the circuit. The logic probe may indicate a 1 or 0, depending on the function of the circuit being tested. Finally, the pushbutton on the logic pulser is activated to place a 1 at the input of the circuit. The LED indicator of the logic probe should

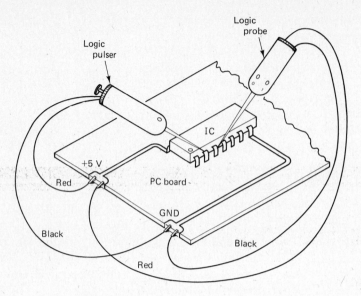

**Figure P-3**  Use of the logic pulser and logic probe.

change states (if 0, it goes to 1, and if 1, it goes to 0) provided that the circuit is functioning properly.

## SECTION B-4
## IC TEST DEVICES

Two other very useful IC test devices for testing digital ICs are shown in Figure P-4. If ICs are mounted close to the PC board and near other components, it is difficult to use logic pulsers and probes to test the circuits. An IC pin extender test clip (Figure P-4a) is placed over the IC and the protruding extender pins make it easier to test the circuit.

The logic monitor clip (Figure P-4b) is actually a more sophisticated logic probe that is placed over a digital IC and gives the state or status of each pin with the use of LEDs simultaneously. It uses diode circuitry that automatically selects the $+V_{CC}$ and Gnd pins, since the location of the power supply pins varies on different function ICs.

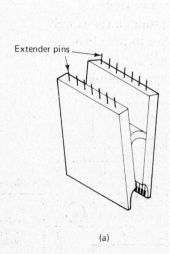
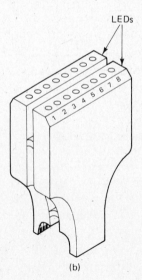

**Figure P-4**  IC test devices: (a) IC pin extender test clip; (b) logic monitor clip. (From F. Hughes, *Illustrated Guidebook to Electonic Devices and Circuits,* Prentice-Hall, Englewood Cliffs, N.J., © 1983, Fig. 14-25, p. 426. Reprinted with permission.)

# Unit C
# Preliminary Review

# Procedure for Testing
# a Digital Circuit

It is sometimes necessary to disconnect a PC board from a digital system to test it separately. The following procedures can be used as a guide for setting up the equipment to check a PC board or for experimenting with a new circuit on a breadboard. Refer to Figure P-5.

1. Have the proper logic diagram (and/or schematic diagram) in front of you.
2. Have the proper power supply, input/output devices, test devices, parts, and test leads in front of you.
3. Construct the circuit if it is an experiment.
4. Connect the power supply to the circuit.
5. Connect all test equipment or indicators to the common circuit ground (as indicated by the dashed lines).
6. Turn on the power supply and using a logic probe, test for the proper indications at the circuit power input connections.
7. Connect the output indicators or logic probe to the circuit.
8. Connect the input switches or logic pulser to the circuit.
9. Set the desired input digital signals (1 or 0).
10. Observe the output device for the correct indication.

Figure P-5  General setup for testing a digital circuit.

7

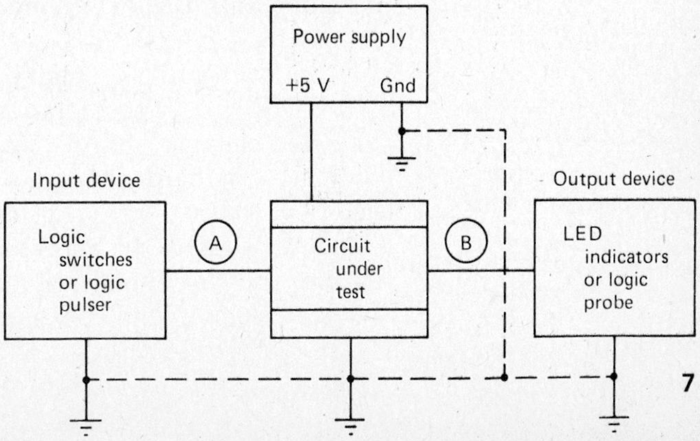

# Unit 1

# Binary Representation and Number Systems

**SECTION 1-1**

**IDENTIFICATION, THEORY, AND OPERATION**

**1-1a  INTRODUCTION: WHY THE USE OF VARIOUS NUMBER SYSTEMS**

Digital circuits and techniques are used in virtually every area of electronic design methods and have provided more practical electronic equipment with increased capabilities and accuracy. Improvements in integrated circuits, particularly digital circuits, have reduced the size, weight, power consumption, and cost of electronic equipment.

There are basically two types of electronic circuits: analog and digital. *Analog circuits*, often referred to as *linear circuits*, deal with continuous values of voltage and current, such as the standard type of audio amplifier. The volume control (a potentiometer) is an analog device also. *Digital circuits* are involved with pulses representing data, control signals, and small increments of measurements. The channel selector on a television receiver uses digital techniques, since each channel is pretuned and a turn of the selector knob brings in a different TV station. However, actual digital circuits use numbers in their operation, such as computers, digital voltmeters, digital thermometers, digital frequency counters, and related equipment. Digital circuits have improved other areas in electronics, such as communications (data transmission), telemetry, industrial controls, and automation.

Electronic systems can be implemented with either the decimal or binary number systems. However, with the use of the *decimal system*, the equipment or hardware must have 10 discrete steps or states representing each of the digits 0 through 9. If the *binary system* is used, only two states, 0 and 1, need to be represented, resulting in much simpler, faster, more reliable, and less expensive equipment. Other number systems (such as octal and hexadecimal) and codes (such as the binary-coded decimal) are multiples of the binary number system and make it easier to communicate and operate with digital circuits. However, regardless of the external number system used with digital equipment, the digital circuits operate internally with numbers of a binary nature.

## 1-1b  BINARY NUMBER SYSTEM

The binary number system has a base or radix of 2, using only the symbols 0 and 1. These *binary* dig*its* are referred to as *bits*. Digital circuits use these bits to represent states or logical conditions. These states can be represented as follows:

| 0 | 1 |
|---|---|
| Low (Lo) | High (Hi) |
| Negative (Gnd) | Positive ($+V$) |
| False | True |
| No | Yes |
| Off | On |
| Open | Closed |

Groups of bits arranged in data form can represent numbers of other systems as shown below.

| Decimal | Binary |
|---------|--------|
| 0 | 0000 |
| 1 | 0001 |
| 2 | 0010 |
| 3 | 0011 |
| 4 | 0100 |
| 5 | 0101 |
| 6 | 0110 |
| 7 | 0111 |
| 8 | 1000 |
| 9 | 1001 |

For your use in converting number systems, it is recommended that the binary equivalents of these numbers be memorized.

## 1-1c  CONVERTING FROM DECIMAL TO BINARY

To convert a decimal number to a binary number, simply divide the decimal number by 2, and the remainder from each division results in the binary number. The first remainder is the *least significant bit* (LSB) and the order increases to the last remainder, which is the *most significant bit* (MSB) of the binary number. For example, convert the decimal number 12 to its equivalent binary number.

**Problem:**                                    **Solution:**

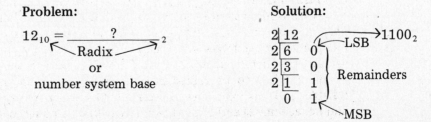

Arrange the binary number on a horizontal line, as shown, indicating its radix. The binary number is not read as eleven hundred, but as one, one, zero, zero, because it is in base 2.

### 1-1d  CONVERTING FROM BINARY TO DECIMAL

All number systems utilize positional notation that is similar to the decimal number system. As the number increases to the left of the decimal point, the value of each position is greater by a factor equal to the base of the number system. With the decimal number system, a number increases by powers of 10, such as units ($10^0$), tens ($10^1$), hundreds ($10^2$), thousands ($10^3$), and so on. For example:

Positional      $10^3$  $10^2$  $10^1$  $10^0$      With each            1000  100  10  1
  notation:   1      3      5      2          power raised:         1      3    5  $2_{10}$

or expressed another way:

$$
\begin{array}{cccc}
1 & 3 & 5 & 2_{10}
\end{array}
$$

                       → $2 \times 10^0$ (1)   =    2
                  → $5 \times 10^1$ (10)  =   50
           → $3 \times 10^2$ (100)  =  300
      → $1 \times 10^3$ (1000) = 1000
                                    $\overline{1352_{10}}$

Positional notation for the binary number system is similar and would be

$2^3$  $2^2$  $2^1$  $2^0$      Or with              8  4  2  1
1      1      0      $0_2$       powers raised:  1  1  0  $0_2$

To convert from a binary number to a decimal number, simply add up the products of the raised powers for each position where a 1 occurs. For example:

**Problem:**                                      **Solution:**

$1100_2 =$ _____?_____ $_{10}$                8  4  2  1
                                       1  1  0  $0_2$
                                                  → $0 \times 1 =$    0
                                       → $0 \times 2 =$    0
                             → $1 \times 4 =$    4
                     → $1 \times 8 = + 8$
                                       $\overline{12_{10}}$

Following is a simple set of rules for converting a binary number to a decimal number:

| Rule | Example |
|---|---|
| 1. Place the raised powers above each bit in the number. | $\longrightarrow$ 8  4  2  1 <br> 1  0  0  $1_2$ |
| 2. Add the power of each position that contains a 1. | 8  4  2  1 <br> 1  0  0  1 |

$$\longrightarrow 1$$
$$\longrightarrow + \underline{8}$$
$$9_{10}$$

## 1-1e  OPERATING WITH BINARY FRACTIONS

The fractional part of a number is to the right of the decimal place and is represented by negative power factors to the base of the number system. The decimal number system uses tenths ($10^{-1} = 1/10 = 0.1$), hundredths ($10^{-2} = 1/100 = 0.01$), thousandths ($10^{-3} = 1/1000 = 0.001$), and so on. For example, $0.75_{10}$ can be written as

$$0.75_{10}$$

$$\longrightarrow 5 \times 10^{-2} \ (0.01) = \quad 0.05$$
$$\longrightarrow 7 \times 10^{-1} \ (0.10) = + \ \underline{0.70}$$
$$0.75_{10}$$

The binary number system fractional numbers are represented similarly as $2^{-1} = 1/2 = 0.5$, $2^{-2} = 1/4 = 0.25$, $2^{-3} = 1/8 = 0.125$, and so on. For example, $0.0101_2$ can be written as

$$0.101$$

$$\longrightarrow 1 \times 2^{-3} \ (0.125) = 0.125$$
$$\longrightarrow 0 \times 2^{-2} \ (0.25) \ = 0.000$$
$$\longrightarrow 1 \times 2^{-1} \ (0.5) \ \ = \underline{0.500}$$
$$0.625_{10}$$

This example also shows the conversion from a binary fraction to a decimal fraction. A simpler form could be used where each equivalent decimal fraction is placed above its respective binary position and then each binary position containing a 1 would have its decimal equivalent added together. For example:

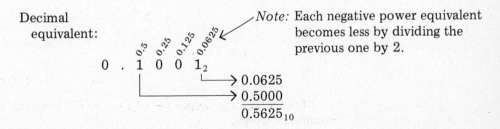

Decimal equivalent:

*Note:* Each negative power equivalent becomes less by dividing the previous one by 2.

$$0 \ . \ 1 \ 0 \ 0 \ 1_2$$

$$\longrightarrow 0.0625$$
$$\longrightarrow \underline{0.5000}$$
$$0.5625_{10}$$

To convert from a decimal fraction to a binary fraction, the opposite operation of multiplying by 2 is used (whole numbers were converted by dividing by 2). In this case, only the fractional part of the decimal number is multiplied by 2, with the resulting carry being the binary equivalent. As an example:

**Problem:**                                                    **Solution:**

$0.625_{10} = \underline{\quad ? \quad}_2$

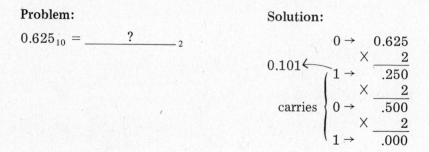

The following example shows the steps in converting a whole and fractional decimal number to its binary equivalent.

**Problem:**

$25.35_{10} = \underline{\quad ? \quad}_2$

**Solution:**

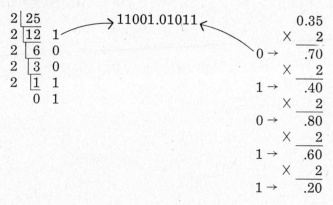

**Proof:**

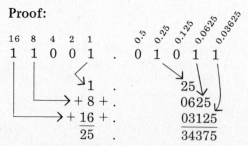

As shown, many binary fractions are unending numbers, and rounding off is often used.

## 1-1f  BINARY REPRESENTATION OF POSITIVE AND NEGATIVE LOGIC

There are two methods of representing binary numbers with voltages. *Positive logic*, shown in Figure 1-1a and b, states that the most negative voltage

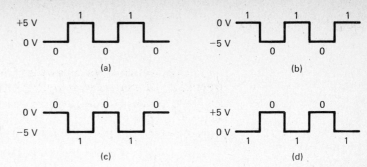

**Figure 1-1** Binary representation with voltages: (a) and (b) positive logic; (c) and (d) negative logic.

represents a 0 and the most positive voltage represents a 1. Most digital circuits use the form shown in Figure 1-1a, where 0 V or ground is equal to 0 and a positive voltage greater than 2 V is equal to 1. The 7400 TTL series and 4000 CMOS series ICs use this form. *Negative logic*, shown in Figure 1-1c and d, states that the most negative voltage represents a 1 and the most positive voltage represents a 0. When digital circuits and equipment have differing positive and negative logic levels, special interface circuits are used to match the various voltage levels.

### 1-1g  THE OCTAL NUMBER SYSTEM

The octal number system uses the numbers 0 through 7 and has a radix of 8. The octal number system is a multiple of the binary system since $2^3 = 8$. Remember that the symbols 8 and 9 are not used in the octal system; therefore, $8_{10} = 10_8$ and $9_{10} = 11_8$.

The same procedures for converting decimal to octal, and vice versa, are used as with converting decimal and binary, except that 8 is used for dividing and the powers of 8 are used in positional notation, respectively. For example:

**Problem:**

$47_{10} = \underline{\quad ? \quad}_8$

**Solution:**

$$8 \underline{|47} \qquad \rightarrow 57_8$$
$$8 \underline{|\ 5\ }\ \ 7$$
$$\ \ \ 0\ \ \ 5$$

**Problem:**

$177_8 = \underline{\quad ? \quad}_{10}$

**Solution:**

$$
\begin{array}{ccc}
(64) & (8) & (1) \\
8^2 & 8^1 & 8^0 \\
1 & 7 & 7
\end{array}
$$

$$7 \times 1 = \quad 7$$
$$7 \times 8 = \quad 56$$
$$1 \times 64 = \quad 64$$
$$\overline{\quad 127_{10}}$$

Converting from octal to binary, and vice versa, is very simple and easy if you have memorized the binary number equivalent of the first eight decimal numbers. To convert from octal to binary, simply write each octal digit in its binary equivalent using three bit places in order. For example:

**Problem:**

$23_8 = \underline{\quad ? \quad}_2$

**Solution:**

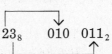

$23_8 \qquad 010 \ \ 011_2$

13

To prove that your answer is correct, convert both numbers to their decimal equivalent, which should be the same number. For example:

**Proof:**

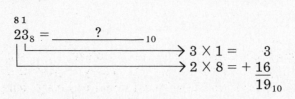

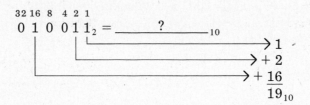

A similar method is used for converting from binary to octal. First arrange the bits into groups of three going from right to left and then convert each group to its octal equivalent. For example:

**Problem:**                                    **Solution:**

$$101111_2 = \underline{\qquad ? \qquad}_8$$

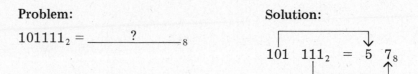

Positional notation can now be used to check the answer, as was done previously.

### 1-1h  THE HEXADECIMAL NUMBER SYSTEM

The radix of the hexadecimal number system is 16, but since there is no singular arithmetic symbol above 9, the alphabetical symbols A through F are used to represent 10 through 15, respectively; that is, $10 = A$, $11 = B$, $12 = C$, $13 = D$, $14 = E$, and $15 = F$. The hexadecimal system is also a multiple of the binary system since $2^4 = 16$.

The same procedures for converting decimal to hexadecimal, and vice versa, are used as with the preceding two number systems, except that 16 is used for dividing and the powers of 16 are used in positional notation, respectively. For example:

**Problem:**                                    **Solution:**

$$47_{10} = \underline{\qquad ? \qquad}_{16}$$

$$
\begin{array}{r}
16 \underline{|47} \\
16 \underline{|\,2} \quad 15 = F \longrightarrow 2F_{16} \\
0 \quad 2
\end{array}
$$

**Problem:**                                    **Solution:**

$A4_{16} = \underline{\qquad ? \qquad}_{10}$

16 1
$A4_{16}$
$\longrightarrow 4 \times 1 \qquad 4$
$\longrightarrow A(10) \times 16 = \underline{160}$
$\overline{164_{10}}$

Converting from hexadecimal to binary, and vice versa, is very simple and easy if you have memorized the binary equivalent of the first 16 decimal numbers. To convert from hexadecimal to binary, simply write each hexadecimal in its binary equivalent using four bit places in order. For example:

**Problem:**                                    **Solution:**

$5B_{16} = \underline{\qquad ? \qquad}_{2}$

$5B_{16} = 0101 \quad 1011_2$

To prove that your answer is correct, convert both numbers to their decimal equivalent, which should be the same number. For example:

**Proof:**

16  1
$5 \; B_{16} = \underline{\qquad ? \qquad}_{10}$

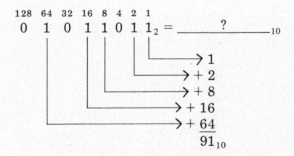

$\longrightarrow 11 \times 1 = \qquad 11$
$\longrightarrow 5 \times 16 = + \underline{80}$
$\overline{91_{10}}$

128  64  32  16  8  4  2  1
$0 \quad 1 \quad 0 \quad 1 \quad 1 \quad 0 \quad 1 \quad 1_2 = \underline{\qquad ? \qquad}_{10}$

$\longrightarrow 1$
$\longrightarrow + 2$
$\longrightarrow + 8$
$\longrightarrow + 16$
$\longrightarrow + \underline{64}$
$\overline{91_{10}}$

A similar method is used for converting from binary to hexadecimal. First arrange the bits into groups of four going from right to left and then convert each group to its hexadecimal equivalent. For example:

**Problem:**                                    **Solution:**

$10011111_2 = \underline{\qquad ? \qquad}_{16}$

$1001 \quad 1111 \quad = \quad 9 \quad F_{16}$

Positional notation can now be used to check the answer.

### 1-1i  THE BINARY-CODED-DECIMAL 8421 CODE

The binary-coded decimal (BCD) 8421 code is not a number system but a code to facilitate the handling of decimal and binary numbers. Once you have memorized the binary equivalent of decimal numbers 0 through 9, conversion between these two systems can be done instantly. However, arithmetic operations using the BCD 8421 code and the required additional circuitry in equipment is a little more complex. The BCD 8421 code uses 4 bits to represent each of the 10 decimal digits and each digit is converted separately. For example:

**Problem:**

$92_{10} = \underline{\qquad ? \qquad}{}_{BCD}$

**Solution:**

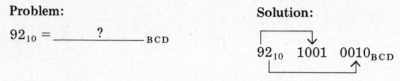

$$92_{10} \quad 1001 \quad 0010_{BCD}$$

The reverse procedure is used to convert from BCD to decimal. For example:

**Problem:**

$0100 \quad 0101_{BCD} = \underline{\qquad ? \qquad}{}_{10}$

**Solution:**

$$0100 \quad 0101 \quad = \quad 4 \quad 5_{10}$$

### 1-1j  COMPARISON OF NUMBER SYSTEMS

Table 1-1 shows a comparison of the number systems covered in the earlier sections. Using this table will help you to recognize quickly the relationship between the various systems and also aid you in memorizing the important equivalent numbers.

**TABLE 1-1**

| Decimal | Binary | Octal | Hexadecimal | BCD 8421 |
|---|---|---|---|---|
| 0 | 0000 | 0 | 0 | 0000 |
| 1 | 0001 | 1 | 1 | 0001 |
| 2 | 0010 | 2 | 2 | 0010 |
| 3 | 0011 | 3 | 3 | 0011 |
| 4 | 0100 | 4 | 4 | 0100 |
| 5 | 0101 | 5 | 5 | 0101 |
| 6 | 0110 | 6 | 6 | 0110 |
| 7 | 0111 | 7 | 7 | 0111 |
| 8 | 1000 | 10 | 8 | 1000 |
| 9 | 1001 | 11 | 9 | 1001 |
| 10 | 1010 | 12 | A | 0001 0000 |
| 11 | 1011 | 13 | B | 0001 0001 |
| 12 | 1100 | 14 | C | 0001 0010 |
| 13 | 1101 | 15 | D | 0001 0011 |
| 14 | 1110 | 16 | E | 0001 0100 |
| 15 | 1111 | 17 | F | 0001 0101 |

**SECTION 1-2**
**EXERCISES I**

Perform all the exercises in this section before beginning the next section.

1. Convert the following decimal numbers to binary numbers; and vice versa to prove your answers (show your work).

   **a.** $6_{10} = $ _____ $_2$          **b.** $9_{10} = $ _____ $_2$          **c.** $15_{10} = $ _____ $_2$

   **d.** $27_{10} = $ _____ $_2$          **e.** $56_{10} = $ _____ $_2$          **f.** $78_{10} = $ _____ $_2$

   **g.** $100_{10} = $ _____ $_2$                    **h.** $125_{10} = $ _____ $_2$

   **i.** $256_{10} = $ _____ $_2$                    **j.** $342_{10} = $ _____ $_2$

**k.** $628_{10} =$ _____ $_2$          **l.** $1024_{10} =$ _____ $_2$

2.  Convert the following binary numbers to decimal numbers; and vice versa to prove your answers (show your work).

**a.** $0111_2 =$ _____ $_{10}$          **b.** $10000_2 =$ _____ $_{10}$

**c.** $1010_2 =$ _____ $_{10}$          **d.** $10010011_2 =$ _____ $_{10}$

**e.** $10011100_2 =$ _____ $_{10}$          **f.** $11001100_2 =$ _____ $_{10}$

    **g.** $10111001_2 =$ _____ $_{10}$         **h.** $11011001_2 =$ _____ $_{10}$

    **i.** $11111001_2 =$ _____ $_{10}$         **j.** $11111111_2 =$ _____ $_{10}$

    **k.** $10010110_2 =$ _____ $_{10}$

**3.** Convert the following decimal fractions to binary fractions; and vice versa to prove your answers (show your work).

    **a.** $0.75_{10} =$ _____ $_2$         **b.** $0.25_{10} =$ _____ $_2$

c. $0.125_{10} = $ _____ $_2$

d. $0.0625_{10} = $ _____ $_2$

e. $0.45_{10} = $ _____ $_2$

f. $0.735_{10} = $ _____ $_2$

4. Convert the following binary fractions to decimal fractions; and vice versa to prove your answers (show your work).

a. $0.101_2 = $ _____ $_{10}$

b. $0.111_2 = $ _____ $_{10}$

c. $0.0001_2 = $ _____ $_{10}$

d. $0.10001_2 = $ _____ $_{10}$

5. Draw a three-pulse square waveform and indicate positive logic where $0 =$ ground and $1 = +5$ V.

6. Convert the following numbers from decimal to octal; and vice versa to prove your answers (show your work).

   a. $17_{10} = $ _____ $_8$                       b. $25_{10} = $ _____ $_8$

   c. $72_{10} = $ _____ $_8$                       d. $125_{10} = $ _____ $_8$

   e. $378_{10} = $ _____ $_8$                      f. $530_{10} = $ _____ $_8$

7. Convert the following numbers as indicated.

    a. $27_8 = $ _____ $_2$             b. $17_8 = $ _____ $_2$

    c. $33_8 = $ _____ $_2$             d. $126_8 = $ _____ $_2$

    e. $234_8 = $ _____ $_2$            f. $101101_2 = $ _____ $_8$

    g. $111010_2 = $ _____ $_8$         h. $100101_2 = $ _____ $_8$

    i. $01011011_2 = $ _____ $_8$       j. $1100101_2 = $ _____ $_8$

8. Convert the following decimal numbers to hexadecimal numbers; and vice versa to prove your answers (show your work).

    a. $16_{10} = $ _____ $_{16}$         b. $27_{10} = $ _____ $_{16}$

    c. $45_{10} = $ _____ $_{16}$         d. $86_{10} = $ _____ $_{16}$

    e. $93_{10} = $ _____ $_{16}$         f. $232_{10} = $ _____ $_{16}$

9. Convert the following numbers as indicated.

    **a.** $34_{16} = $ _____ $_2$       **b.** $58_{16} = $ _____ $_2$

    **c.** $96_{16} = $ _____ $_2$       **d.** $117_{16} = $ _____ $_2$

    **e.** $256_{16} = $ _____ $_2$       **f.** $00101001_2 = $ _____ $_{16}$

    **g.** $01000111_2 = $ _____ $_{16}$       **h.** $10000001_2 = $ _____ $_{16}$

    **i.** $10100100_2 = $ _____ $_{16}$       **j.** $11111001_2 = $ _____ $_{16}$

10. Convert the following numbers as indicated.

    **a.** $27_{10} = $ _____ BCD       **b.** $56_{10} = $ _____ BCD

    **c.** $93_{10} = $ _____ BCD       **d.** $125_{10} = $ _____ BCD

    **e.** $435_{10} = $ _____ BCD       **f.** $00110101_{BCD} = $ _____ $_{10}$

    **g.** $10000110_{BCD} = $ _____ $_{10}$       **h.** $10011001_{BCD} = $ _____ $_{10}$

    **i.** $00010101_{BCD} = $ _____ $_{10}$       **j.** $01111000_{BCD} = $ _____ $_{10}$

## SECTION 1-3
## DEFINITION EXERCISES

Give a brief description of each of the following number systems and terms.
State the numbers used in each system.

1. Decimal _____

_____

_____

2. Binary _____

_____

_____

3. Octal _____

_____

_____

4. Hexadecimal _____

_____

_____

**5.** BCD 8421 code _____

_____

_____

**6.** Positive logic _____

_____

_____

**7.** Negative logic _____

_____

_____

**8.** Bit _____

_____

_____

### SECTION 1-4
### EXPERIMENTS

This experiment can be performed on a standard logic trainer or be constructed from discrete components (see the appendix).

### EXPERIMENT 1. IDENTIFYING EQUIVALENT BINARY/DECIMAL NUMBERS

*Objective:*

To show the equivalent binary representation of decimal numbers 0 through 15.

*Introduction:*

Four switches are wired to four LEDs and are properly assigned binary-weighted values as shown in Figure 1-2. The switches are then set according to the table shown in Figure 1-2c. The values of the "on" switches are then added together to produce the equivalent decimal number.

*Materials Needed:*

*1*    Digital logic trainer, *or*
*1*    +5-V regulated power supply
*1*    Breadboard for constructing circuit
*4*    SPDT switches
*4*    LEDs at $V_F \approx 2$ V
*4*    220-$\Omega$ resistors at 0.5 W
     Several hookup wires or leads

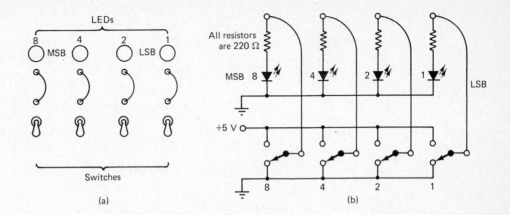

Figure 1-2 Equivalent binary/decimal numbers: (a) logic trainer connections; (b) discrete component wiring; (c) conversion table.

| Binary number | | | | LEDs on | Decimal number |
|---|---|---|---|---|---|
| 8 | 4 | 2 | 1 | | |
| 0 | 0 | 0 | 0 | | |
| 0 | 0 | 0 | 1 | | |
| 0 | 0 | 1 | 0 | | |
| 0 | 0 | 1 | 1 | | |
| 0 | 1 | 0 | 0 | | |
| 0 | 1 | 0 | 1 | | |
| 0 | 1 | 1 | 0 | | |
| 0 | 1 | 1 | 1 | | |
| 1 | 0 | 0 | 0 | | |
| 1 | 0 | 0 | 1 | | |
| 1 | 0 | 1 | 0 | | |
| 1 | 0 | 1 | 1 | | |
| 1 | 1 | 0 | 0 | | |
| 1 | 1 | 0 | 1 | | |
| 1 | 1 | 1 | 0 | | |
| 1 | 1 | 1 | 1 | | |

(c)

*Procedure:*

1. Connect the circuit shown in Figure 1-2a (or 1-2b).
2. Set the four switches to the positions shown in the four columns under the heading "Binary number." Let a high (or up) switch = 1 and a low (or down) switch = 0.
3. List the positions of the lighted LEDs in the column marked "LEDs on."
4. Add up the weighted positions in the "LEDs on" column and write the equivalent decimal number in the "Decimal number" column.
5. Follow steps 1 to 4 for all binary numbers.

*Fill-In Questions:*

1. Moving from right to left in a binary-weighted number, each position's value is multiplied by _____ .

2. In a binary number, if positions 1 and 8 are a 1, the decimal equivalent number is _____ .

3. To show the binary equivalent of decimal number 15, weighted positions _____ will have to be a 1.

## SECTION 1-5
## INSTANT REVIEW

- The base or radix of a number system indicates how many symbols are used within it.
- The binary number system has a radix of 2 with 0 and 1 as its symbols. To convert decimal whole numbers to binary, you divide by 2, with the remainders becoming the answer. To convert decimal fractional numbers to binary, you multiply by 2 and the carries become the answer.
- The octal number system has a radix of 8 with numbers 0 through 7 as its symbols. Decimal-to-octal conversion is the same procedure as with binary, except that the number 8 is used. The octal number system is a multiple of the binary system since $2^3 = 8$ and conversion between these two systems is easier.
- The hexadecimal number system has a radix of 16 with numbers 0 through 9 and characters A through F as its symbols. Decimal-to-hexadecimal conversion is the same procedure as with binary except that the number 16 is used. The hexadecimal number system is a multiple of the binary system since $2^4 = 16$ and conversion between these two systems is easier.
- The BCD 8421 code is not a number system but uses 4 bits to represent all decimal numbers, and conversion between the two systems is much easier. Positional notation with the BCD 8421 code will not produce the same equivalent decimal number as the other systems.
- Positive logic uses the more positive voltage as a 1, and negative logic uses the more negative voltage as 1.

## SECTION 1-6
## EXERCISES II

Perform all the exercises in this section before beginning the next section.

Convert the following numbers to the other number systems as indicated. In some cases you may have to convert to decimal first, to find other conversions.

1. $36_{10} =$ _____ $_2 =$ _____ $_8 =$ _____ $_{16} =$ _____ BCD

2. $159_{10} =$ _____$_2 =$ _____$_8 =$ _____$_{16} =$ _____BCD

3. $11111010_2 =$ _____$_{10} =$ _____$_8 =$ _____$_{16} =$ _____BCD

4. $10010111_2 =$ _____$_{10} =$ _____$_8 =$ _____$_{16} =$ _____BCD

5. $57_8 =$ _____$_2 =$ _____$_{10} =$ _____$_{16} =$ _____BCD

6. $127_8 =$ _____$_2 =$ _____$_{10} =$ _____$_{16} =$ _____BCD

7. $5A_{16} =$ _____$_2 =$ _____$_{10} =$ _____$_8 =$ _____BCD

8. $E7_{16} =$ _____$_2 =$ _____$_{10} =$ _____$_8 =$ _____BCD

9. $01110101_{BCD} =$ _____$_{10} =$ _____$_2 =$ _____$_8 =$ _____$_{16}$

10. $10011001_{BCD} =$ _____$_{10} =$ _____$_2 =$ _____$_8 =$ _____$_{16}$

## SECTION 1-7
## TROUBLESHOOTING APPLICATION: NUMBER CONVERSION

In the following conversions there are some mistakes. Circle the ones that are

not correct and write in the correct answer. Example: (a.)  $12_{10} = \overset{1100}{\cancel{1011}}_2$

1. $24_{10} = 11000_2$

2. $33_{10} = 111001_2$

3. $157_{10} = 10011101_2$

4. $10101_2 = 42_{10}$

5. $100110_2 = 38_{10}$

6. $10011111_2 = 159_{10}$

7. $10_{10} = 10_8$

8. $24_8 = 13_{10}$

9. $123_{10} = 173_8$

10. $43_8 = 33_{10}$

11. $177_8 = 127_{10}$

12. $30_8 = 27_{10}$

13. $110101_2 = 56_8$

14. $100011_2 = 43_8$

15. $1001011_2 = 113_{10}$

16. $23_8 = 010110_2$

17. $47_8 = 100111_2$

18. $54_8 = 101100_2$

19. $18_{10} = 12_{16}$

20. $62_{10} = 3C_{16}$

21. $242_{10} = F2_{16}$

22. $12F_{16} = 303_{10}$

23. $BE_{16} = 180_{10}$

24. $2F_{16} = 47_{10}$

25. $10010111_2 = A7_{16}$

26. $10111001_2 = C9_{16}$

27. $11110000_2 = F0_{16}$

28. $6F_{16} = 01101110_2$

29. $89_{16} = 10001001_2$

30. $3C_{16} = 00111011_2$

31. $45_{10} = 01000101_{BCD}$

32. $70_{10} = 00111000_{BCD}$

33. $96_{10} = 01101001_{BCD}$

34. $10000011_{BCD} = 83_{10}$

35. $00100001_{BCD} = 21_{10}$

36. $01001001_{BCD} = 48_{10}$

**SECTION 1-8**
**SELF-CHECKING QUIZZES**

**1-8a   NUMBER SYSTEMS: TRUE-FALSE QUIZ**

Place a T for true or an F for false to the left of each statement.

_____ 1. The base or radix of a number system indicates how many symbols the system uses.

_____ 2. The symbols used for the octal system are 0 through 8.

_____ 3. The BCD 8421 code can be converted directly to any number system.

_____ 4. The hexadecimal number system uses the numbers 0 through 9 and the alphabetical letters A through F.

_____ 5. The process of converting binary to octal involves using the bits in groups of three.

_____ 6. The octal and hexadecimal number systems are multiples of the binary number system.

_____ 7. To convert a whole decimal number to a hexadecimal number, you divide by 4 and the remainders become the answer.

_____ 8. The BCD 8421 code and the binary number system can be grouped into bits of 4 for converting.

_____ 9. To convert a fractional decimal number into a binary number, you multiply by 2 and the carries become the answer.

_____ 10. With positive logic $0V = 1$ and $+5 V = 0$.

**1-8b   NUMBER SYSTEMS: MULTIPLE-CHOICE QUIZ**

Circle the correct answer for each question.

1. The equivalent binary number for $101.625_{10}$ is:
   a. $1000111.110_2$          b. $1100101.101_2$
   c. $1110001.011_2$          d. $1001111.101_2$

2. The equivalent decimal number for $10110111.011_2$ is:
   a. $14.45_{10}$             b. $79.30_{10}$
   c. $93.275_{10}$            d. $183.375_{10}$

3. The equivalent octal number for $1101101_2$ is:
   a. $17_8$                   b. $36_8$
   c. $155_8$                  d. $1231_8$

4. The equivalent BCD 8421 code number for $945_{10}$ is:

   a. $1110110001_{BCD}$  b. $1001100101_{BCD}$

   c. $100101000101_{BCD}$  d. $111101001010_{BCD}$

5. The pulse train shown in Figure 1-3 can be represented as:

   a. $100101_2$  b. $101010_2$

   c. $010101_2$  d. $011010_2$

Figure 1-3  +5 V  0 V

6. The equivalent hexadecimal number for $108_{10}$ is:

   a. $27_{16}$  b. $54_{16}$

   c. $A8_{16}$  d. $6C_{16}$

7. The equivalent binary number for $4F_{16}$ is:

   a. $01001111_2$  b. $10011_2$

   c. $10001101_2$  d. $11110100_2$

8. The equivalent decimal number for $A2C_{16}$ is:

   a. $2604_{10}$  b. $3712_{10}$

   c. $92_{10}$  d. $24_{10}$

9. The equivalent hexadecimal number for $10011101_2$ is:

   a. $157_{16}$  b. $253_{16}$

   c. $9D_{16}$  d. $E9_{16}$

10. The equivalent BCD 8421 code number for $F9_{16}$ is:

    a. $11111001_{BCD}$  b. $01001111_{BCD}$

    c. $001001001001_{BCD}$  d. $10011010_{BCD}$

## ANSWERS TO EXPERIMENTS AND QUIZZES FOR UNIT 1

*Experiment 1:*

(1) 2  (2) 9  (3) 8421 or 1248

*True-False:*

(1) T  (2) F  (3) F  (4) T  (5) T  (6) T  (7) F  (8) T  (9) T
(10) F

*Multiple-Choice:*

(1) b  (2) d  (3) c  (4) c  (5) b  (6) d  (7) a  (8) a  (9) c
(10) c

# Unit 2

## Logic Gates

**SECTION 2-1**
**IDENTIFICATION, THEORY, AND OPERATION**

### 2-1a  INTRODUCTION

Logic gates are controlled electronic switches. However, unlike switches, they allow data to pass in only one direction—from input to output. The electronic pulses placed at two or more inputs to a logic gate determine the condition at the output. The specific combination of pulses at the inputs that result in a voltage pulse at the output are determined by the logical function of the gate. The logical functions can be explained by Boolean (or logic) algebra.

Components such as resistors, diodes, bipolar transistors, and MOSFETs can be used to construct logic gates. However, most logic gates are in integrated circuit (IC) form. The technician is unable to test and repair these discrete components within the IC. Therefore, only the logic operation of an IC is necessary to understand when troubleshooting digital equipment. Usually, entire IC packages are replaced when any portion is suspected of being defective.

Most logic gate symbols are standardized using military standard 806 (MIL-STD-806) introduced by the U.S. government. This manual uses the standard form and refers to the 7400 TTL and 4000 CMOS series ICs.

### 2-1b  THE AND GATE

The *AND function* states that if all inputs are 1, the output will also be a 1. The switch analogy for the AND operation is shown in Figure 2-1a. In this case, an open switch represents 0 and a closed switch represents a 1. When switch *A* is closed and switch *B* is closed, the LED, representing the output

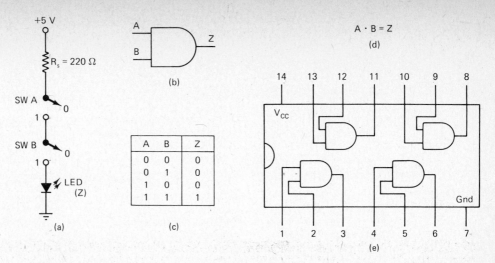

**Figure 2-1** AND gate: (a) switch analogy; (b) logic symbol; (c) truth table; (d) Boolean expression; (e) typical 7408 DIP IC.

($Z$), will turn on (or glow), indicating a 1. When the LED is off, it indicates an output of 0. Figure 2-1b shows the standard logic symbol for an AND gate. The truth table in Figure 2-1c states that when input $A$ is a 1 and input $B$ is a 1, the output ($Z$) will be a 1. Any other combination of input conditions will not produce a 1 at the output. The Boolean expression in Figure 2-1d states that "$A$ AND $B$ equals $Z$." The symbol for multiplication ($\cdot$) stands for AND in Boolean algebra. Figure 2-1e illustrates a typical 7408 quad two-input AND gate DIP IC (dual-in-line-package integrated circuit).

## 2-1c THE OR GATE

The *OR function* states that if any input is a 1, the output will also be a 1. The switch analogy for the OR operation is shown in Figure 2-2a. An open switch represents a 0 and a closed switch represents a 1. When switch $A$ or switch $B$ is closed, the LED ($Z$) will turn on, indicating a 1 at the output. When the LED is off, it indicates an output of 0. Figure 2-2b shows the standard logic symbol for an OR gate. The truth table in Figure 2-2c states that when either input $A$, or input $B$, or both inputs are a 1, the output will be a 1. Since a 1 is also produced at the output when both inputs are 1, this type of gate is referred to as an *inclusive-OR gate*. The Boolean expression in Figure 2-2d states that "$A$ OR $B$ equals $Z$." The symbol for addition (+) stands for OR in Boolean algebra. Figure 2-2e illustrates a typical 7432 quad two-input OR gate DIP IC.

**Figure 2-2** Inclusive-OR gate: (a) switch analogy; (b) logic symbol; (c) truth table; (d) Boolean expression; (e) typical 7432 DIP IC.

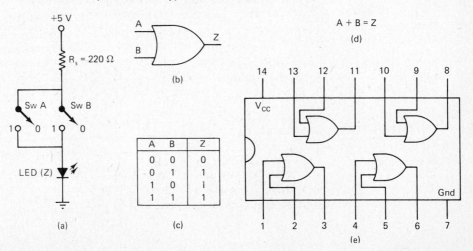

## 2-1d THE BUFFER/DRIVER AND INVERTER

The buffer/driver and inverter are single-input devices that are not considered logic gates, but play extremely important roles in digital electronics. The *buffer/driver switch* analogy shown in Figure 2-3a simply shows that switch A in the open position indicates 0 and the LED is off, but a closed switch indicates 1 and the LED is on. The buffer/driver is represented by a triangle as shown in Figure 2-3b and indicates that the output is the same as the input. (See Figure 2-3c.) This noninverting gate is used to drive or provide input power to other gates and output-display devices or to serve as a buffer in isolating different circuits.

The *inverter function* simply states that the output is the *complement* or *opposite* of its input condition. Its switch analogy is shown in Figure 2-3d. In this case, an open switch represents a 1 and a closed switch represents a 0. Notice that the switch is connected between the cathode of the LED and ground. Ground usually represents a logic 0 in a positive logic system. When switch A is open, the LED is off, indicating an output of 0. When the switch is closed, the LED is on, indicating an output of 1. The standard logic symbol for an inverter has a triangle with a bubble at the apex or output as shown in Figure 2-3e. However, the bubble may be drawn at the input, but the function remains the same. The truth table shown in Figure 2-3f states that when the input A is 0, the output is 1, and when the input is 1, the output is 0. The Boolean expression in Figure 2-3g states "A equals $Z(\overline{A})$." The bar over the A represents "NOT," or the complement of A. A typical 7404 hex inverter DIP IC is shown in Figure 2-3h.

**Figure 2-3** Driver and inverter: (a) driver switch analogy; (b) driver logic symbol; (c) driver Boolean expression; (d) inverter switch analogy; (e) inverter logic symbol; (f) inverter truth table; (g) inverter Boolean expression; (h) typical 7404 DIP IC.

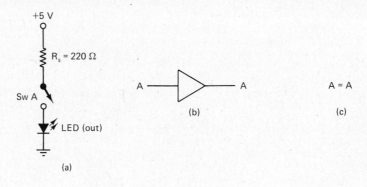

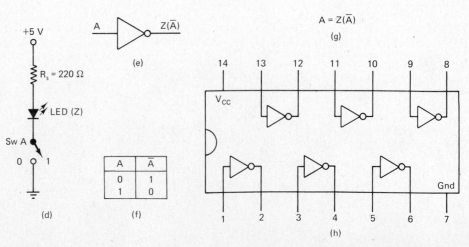

## 2-1e  NEGATED-INPUT GATES

Some logic gates require a combination of 1 and 0 at the inputs in order to turn on. An AND gate with an inverter at one input as shown in Figure 2-4b is a negated input gate. The standard logic symbol usually used is without the triangle and only the bubble represents the inverter, as shown in Figure 2-4c. The switch analogy shown in Figure 2-4a uses both forms of identifying 1 and 0. Switch $A$ in the open position represents a 0, and when it is closed a 1, since it applies a positive voltage to the anode of the LED. Switch $B$ on the cathode side of the LED represents a 1 in the open position and a 0 in the closed position, since it connects the cathode to ground. When both switches are closed, current flows through the LED, causing it to glow, and gives an output indication of 1. The truth table in Figure 2-4d indicates that the output of the inhibit gate will be 1 when input $A$ is 1 and input $B$ is 0. The Boolean expression in Figure 2-4e states that "$A$ and NOT $B$ equals $Z$."

An OR gate with an inverter might be used as shown in Figure 2-4f. The output will be a 1 when input $A$ is 1 or input $B$ is 0.

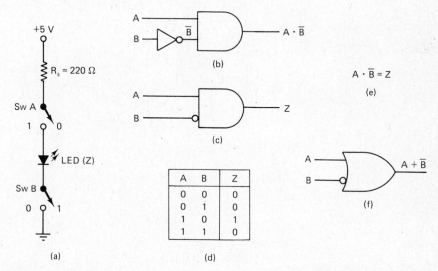

**Figure 2-4**  Negated-input gate: (a) switch analogy; (b) AND gate—B input inverted; (c) logic symbol; (d) truth table; (e) Boolean expression; (f) OR gate with negated input.

## 2-1f  THE EXCLUSIVE-OR GATE

The *exclusive-OR function* states that any single input being a 1 will produce a 1 at the output. If more than one input is 1 at the same time, the output will be 0. The switch analogy shown in Figure 2-5a uses two double-pole, double-throw (DPDT) switches as the inputs. The switches are shown in the 0 position and the LED is off. If only one of either switch is activated (becomes 1), the LED will turn on. If both switches are activated (become 1), the LED will turn off.

An exclusive-OR gate can be formed by connecting negated-input AND gates to an inclusive-OR gate as shown in Figure 2-5b. Notice that inputs $A$ and $B$ are cross-connected to the AND gates. The standard logic symbol for an exclusive-OR gate appears in Figure 2-5c. It is similar to a standard OR gate logic symbol except with a line across the inputs. The truth table shown in Figure 2-5d states that the output will be a 1 only when a single input is 1. The Boolean expression shown in Figure 2-5e states, "$A$ and NOT $B$ or NOT

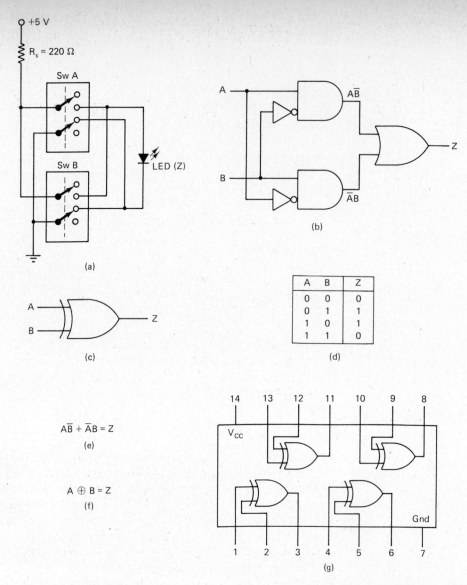

**Figure 2-5** Exclusive-OR gate: (a) switch analogy; (b) separate gates to form function; (c) standard logic symbol; (d) truth table; (e) Boolean expression; (f) simplified expression; (g) typical 7486 DIP IC.

$A$ and $B$ equals $Z$." A simpler and commonly used formula for this gate is shown in Figure 2-5f, with a circle around the OR (+) symbol. A typical 7486 quad two-input exclusive-OR gate DIP IC is shown in Figure 2-5g.

## 2-1g THE NAND GATE

The *NAND function* states that when any or all inputs are 0, the output will produce a 1. The switch analogy for the NAND gate is shown in Figure 2-6a. With this circuit the switches are in parallel and connected to the cathode lead of the LED. An open switch indicates a 1 and a closed switch represents a 0. A closed switch connects the LED to ground, causing it to turn on and indicate a 1 output. Either or both switches closed will turn on the LED. Figure 2-6b shows an AND gate followed by an inverter and may read "NOT AND"—hence the term NAND. The standard symbol for the NAND gate, shown in Figure 2-6c, omits the triangle of the inverter and uses only the bubble at the output. The truth table of Figure 2-6d shows that a 1 will appear at the output of the NAND gate any time that a 0 is present at one or both of the inputs. The Boolean expression in Figure 2-6e reads "$A$ and $B$, NOT" or it can be read as "NOT $A$ and $B$." The formula in Figure 2-6f is

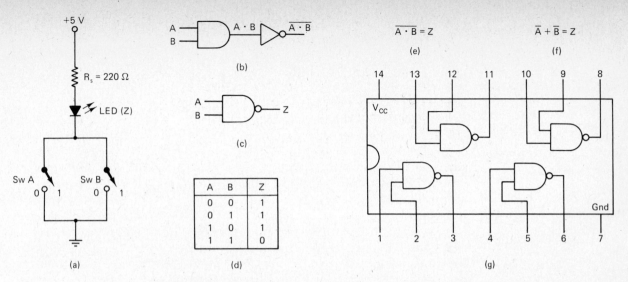

**Figure 2-6** NAND gate: (a) switch analogy; (b) AND-inverter; (c) logic symbol; (d) truth table; (e) Boolean expression; (f) De Morgan expression; (g) typical 7400 DIP IC.

identical to the previous formula and will be explained further in the next section. A NAND gate is similar to an OR gate except that it requires a 0 at any input to turn it on. Figure 2-6g shows a typical 7400 quad two-input NAND gate DIP IC.

### 2-1h THE NOR GATE

The *NOR function* states that when all inputs are 0, the output will be 1. The switch analogy for a NOR gate is shown in Fig. 2-7a. With this circuit the switches are in series and connected to the cathode lead of the LED. Again, an open switch indicates a 1 and a closed switch represents a 0. When both switches are closed, the LED is connected to ground, which turns it on and represents a 1 output. Figure 2-7b shows an OR gate followed by an inverter and may read "NOT OR"—hence the term NOR. The standard logic symbol for the NOR gate, shown in Figure 2-7c, omits the triangle of the inverter and uses only the bubble at the output. The truth table of Figure 2-7d shows that a 1 will appear at the output of the NOR gate when all inputs are 0. The Boolean expression in Figure 2-7e reads "*A* or *B* NOT" or can be read "NOT, *A* or *B*." The formula in Figure 2-7f is identical to the previous formula and will be explained further in the next section. A NOR gate is similar to an AND gate except that all inputs must be 0 to turn it on. Figure 2-7g shows a typical 7402 quad two-input NOR gate DIP IC.

**Figure 2-7** NOR gate: (a) switch analogy; (b) OR-inverter; (c) logic symbol; (d) truth table; (e) Boolean expression; (f) De Morgan expression; (g) typical 7402 DIP IC.

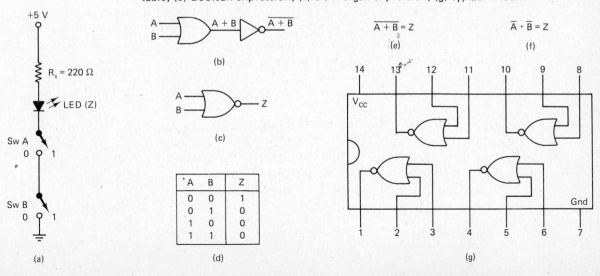

## 2-1i  THE EXCLUSIVE-NOR GATE

The *exclusive-NOR function* states that when all inputs are the same (either all 0 or all 1), the output will produce a 1. The switch analogy shown in Figure 2-8a uses two double-pole, double-throw (DPDT) switches for the inputs, similar to the exclusive-OR gate. The switches are shown in the 0 position and the LED is on, indicating a 1 output. If only one of either switch is activated (becomes 1), the LED will turn off, indicating an output of 0. When both switches are activated (become 1), the LED will turn on.

An exclusive-NOR gate can be formed by connecting two negated-input AND gates with an inclusive-OR gate to produce an exclusive-OR gate and then connecting an inverter to the final output as shown in Figure 2-8b. The standard logic symbol for an exclusive-NOR gate is the same as a standard NOR gate, except for a line across the inputs as shown in Figure 2-8c. The truth table shown in Figure 2-8d indicates that the output will be 1 when both inputs are the same. The Boolean expression in Figure 2-8e states "*A* and *B* or NOT *A* and NOT *B* equals *Z*." For simplicity, the formula in Figure 2-8f can be used for the exclusive-NOR gate. This circuit is then a basic comparator; when one input condition matches the other input condition, the output gives a 1 indication. Figure 2-8g simply states "when *A* equals *B*, the output will be 1."

**Figure 2-8**  Exclusive-NOR gate: (a) switch analogy; (b) separate gates to form function; (c) standard logic symbol; (d) truth table; (e) Boolean expression; (f) simplified expression; (g) comparator statement.

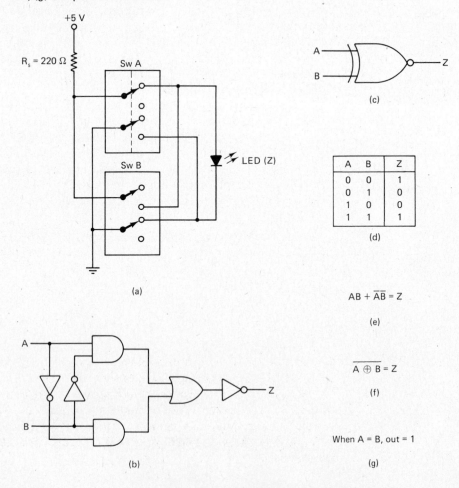

| A | B | Z |
|---|---|---|
| 0 | 0 | 1 |
| 0 | 1 | 0 |
| 1 | 0 | 0 |
| 1 | 1 | 1 |

(d)

$$AB + \overline{A}\,\overline{B} = Z$$

(e)

$$\overline{A \oplus B} = Z$$

(f)

When A = B, out = 1

(g)

## 2-1j  EQUIVALENT GATES

Various logic gates have their *equivalent gates* that perform the same function, but are represented differently, as shown in Figure 2-9. A NAND gate may be represented as a negated-input OR gate (Figure 2-9a). A NOR gate may be shown as a negated-input AND gate (Figure 2-9b). If the NAND gate and NOR gate inputs are connected together, the function becomes an inverter (Figure 2-9c and d). An exclusive-OR gate can be made from two negated-input AND gates and an OR gate (Figure 2-9e). If an inverter is connected to the output of the exclusive-OR gate, it becomes an exclusive-NOR gate (Figure 2-9f).

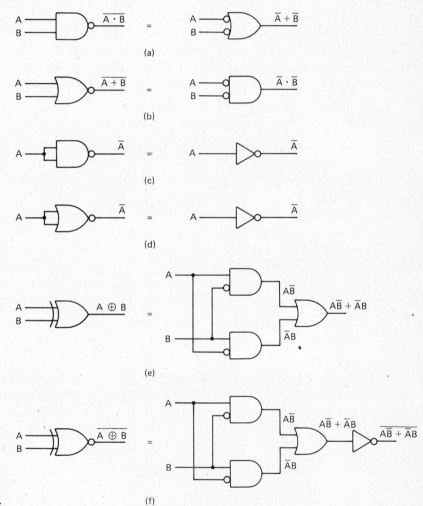

**Figure 2-9**  Equivalent gates: (a) NAND; (b) NOR; (c) NAND inverter; (d) NOR inverter; (e) exclusive-OR; (f) exclusive-NOR.

## 2-1k  EXPANDING GATE INPUTS

The inputs to an AND function can be expanded by connecting the output of one AND gate to an input of another AND gate, as shown in Figure 2-10a. The inputs to an OR function can be expanded by connecting the output of one OR gate to an input of another OR gate, as shown in Figure 2-10b. Usually, this is the only reason for these types of connections, since, normally, AND gates feed OR gates and OR gates feed AND gates in a logical arrangement.

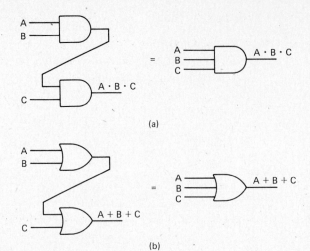

(a)

(b)

**Figure 2-10** Expanding gate inputs: (a) three-input AND function; (b) three-input OR function.

*It is important for the reader to memorize each logic gate symbol, its truth table, its Boolean expression, and be able to state simply how to turn it on so that a 1 is at the output.*

**SECTION 2-2**
**EXERCISES I**

Perform all the exercises in this section before beginning the next section.

1.  Draw the logic symbol, truth table, and Boolean expression for each of the following gates. Label inputs and output with alphabetical letters.

    **a.  AND gate**        **b.  OR gate**        **c.  Inverter**

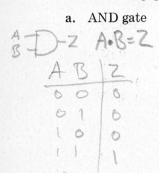

$A \cdot B = Z$

| A | B | Z |
|---|---|---|
| 0 | 0 | 0 |
| 0 | 1 | 0 |
| 1 | 0 | 0 |
| 1 | 1 | 1 |

$A + B = Z$

| A | B | Z |
|---|---|---|
| 0 | 0 | 0 |
| 0 | 1 | 1 |
| 1 | 0 | 1 |
| 1 | 1 | 1 |

$\overline{A} = Z$

| A | Z |
|---|---|
| 0 | 1 |
| 1 | 0 |

    **d.  NAND gate**        **e.  NOR gate**        **f.  Exclusive-OR gate**

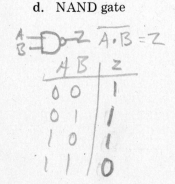

$\overline{A \cdot B} = Z$

| A | B | Z |
|---|---|---|
| 0 | 0 | 1 |
| 0 | 1 | 1 |
| 1 | 0 | 1 |
| 1 | 1 | 0 |

$\overline{A + B} = Z$

| A | B | Z |
|---|---|---|
| 0 | 0 | 1 |
| 0 | 1 | 0 |
| 1 | 0 | 0 |
| 1 | 1 | 0 |

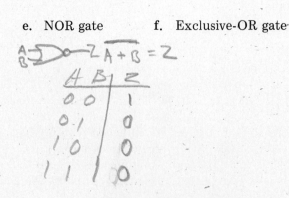

$\overline{A} \cdot B$

**g.** Exclusive-NOR gate        **h.** Inhibit gate        **i.** Noninverting driver

2. Match the logic gates given in column A with their proper truth table
   given under the choices.

|  | *Column A* | | | *Choices* |
|---|---|---|---|---|

*C*  ___ 1.  AND gate

*E*  ___ 2.  OR gate

___ 3.  Driver

*B*  ___ 4.  Inverter

*A*  ___ 5.  NAND gate

*H*  ___ 6.  NOR gate

___ 7.  Exclusive-OR gate

___ 8.  Exclusive-NOR gate

___ 9.  Inhibit gate

**a.**

| A | B | Z |
|---|---|---|
| 0 | 0 | 1 |
| 0 | 1 | 1 |
| 1 | 0 | 1 |
| 1 | 1 | 0 |

**b.**

| A | Z ($\overline{A}$) |
|---|---|
| 0 | 1 |
| 1 | 0 |

**c.**

| A | B | Z |
|---|---|---|
| 0 | 0 | 0 |
| 0 | 1 | 0 |
| 1 | 0 | 0 |
| 1 | 1 | 1 |

**d.**

| A | B | Z |
|---|---|---|
| 0 | 0 | 1 |
| 0 | 1 | 0 |
| 1 | 0 | 0 |
| 1 | 1 | 1 |

**e.**

| A | Z |
|---|---|
| 0 | 0 |
| 1 | 1 |

**f.**

| A | B | Z |
|---|---|---|
| 0 | 0 | 0 |
| 0 | 1 | 1 |
| 1 | 0 | 1 |
| 1 | 1 | 1 |

**g.**

| A | B | Z |
|---|---|---|
| 0 | 0 | 0 |
| 0 | 1 | 0 |
| 1 | 0 | 1 |
| 1 | 1 | 0 |

**h.**

| A+B | Z |
|---|---|
| 0 0 | 1 |
| 0 1 | 0 |
| 1 0 | 0 |
| 1 1 | 0 |

**i.**

| A | B | Z |
|---|---|---|
| 0 | 0 | 0 |
| 0 | 1 | 1 |
| 1 | 0 | 1 |
| 1 | 1 | 0 |

3. Draw an equivalent circuit of the following gates.

   **a.** NAND            **b.** NOR            **c.** Exclusive-OR

d.  Exclusive-NOR      e.  Inverter

4.  Match the logic symbols in column A to their respective formula in column B.

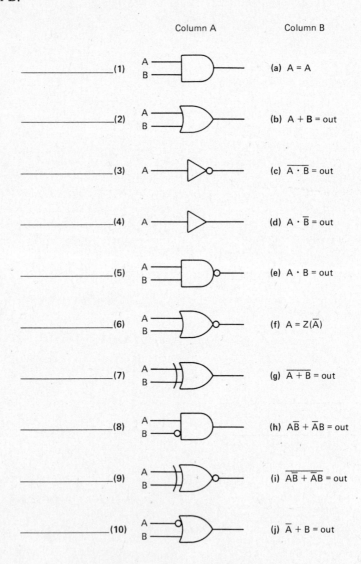

Column A                    Column B

_____(1)    A ──┐D──    (a)  A = A
                 B ──┘

_____(2)    A ──┐D──    (b)  A + B = out
                 B ──┘

_____(3)    A ──▷∘──    (c)  $\overline{A \cdot B}$ = out

_____(4)    A ──▷──     (d)  $A \cdot \overline{B}$ = out

_____(5)    A ──┐D∘──   (e)  A · B = out
                 B ──┘

_____(6)    A ──┐D∘──   (f)  A = Z($\overline{A}$)
                 B ──┘

_____(7)    A ──┐)D──   (g)  $\overline{A + B}$ = out
                 B ──┘

_____(8)    A ──┐D──    (h)  $A\overline{B} + \overline{A}B$ = out
                 B ──∘┘

_____(9)    A ──┐)D∘──  (i)  $\overline{A\overline{B} + \overline{A}B}$ = out
                 B ──┘

_____(10)   A ──∘┐D──   (j)  $\overline{A} + B$ = out
                 B ──┘

5. Draw a method of expanding the number of inputs to an AND gate.

6. Draw a method of expanding the number of inputs to an OR gate.

**SECTION 2-3**
**DEFINITION EXERCISES**

Give a brief description of the operation for each of the following logic gates. Use 1s and 0s for conditions at the inputs and outputs.

1. AND gate _____

_____

_____

2. OR gate _____

_____

_____

3. Inverter _____

_____

_____

4. Driver _____

_____

_____

5. NAND gate _____

_____

_____

6. NOR gate _____

_____

_____

7. Inhibit gate _____

_____

_____

8. Exclusive-OR gate _____

_____

_____

9. Exclusive-NOR gate _____

_____

_____

## SECTION 2-4
## EXPERIMENTS

The experiments in this section can be performed on most standard commercial logic trainers.

The same procedures will be used for each of the experiments in this section. Two switches are wired to inputs $A$ and $B$ of the gate (except for the inverter). A lamp or indicator such as an LED is wired to the output. The truth table is then used to set the switches for the various conditions at the inputs, while the output is monitored for verification. A 0 means that the switch is down and a 1 means that the switch is up. The output will indicate 0 when the light is off and 1 when the light is on.

## EXPERIMENT 1.   THE AND GATE

*Objective:*

To demonstrate the operation of an AND gate and verify its truth table.

*Introduction:*

The AND gate will produce a 1 at the output when all inputs are 1.

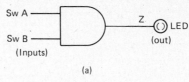

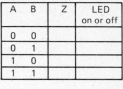

(a)

**Figure 2-11**  AND gate: (a) logic symbol; (b) Boolean expression; (c) truth table.

$A \cdot B = Z$

(b)

(c)

*Materials Needed:*

1  Digital logic trainer, *or*

1  7408 quad two-input AND gate DIP IC

1  Light indicator

2  Input switches

Several hookup wires or leads

*Procedure:*

1.  Connect the AND gate as shown in Figure 2-11a.
2.  Set the input switches according to the truth table as shown in the first row.
3.  Monitor the output for the proper indication. If the light is off, place a 0 under column Z and write "off" in the LED column. If the light is on, place a 1 under column Z and write "on" in the LED column.
4.  Repeat steps 2 and 3 for the remaining rows of the truth table.

*Fill-In Questions:*

1.  The AND gate turns on when _____ inputs are 1.

2.  If one input is a 1 and the other input is a 0, the output for the AND gate will be _____.

3.  The Boolean expression for the AND gate is _____.

**EXPERIMENT 2.  THE OR GATE**

*Objective:*

To show the operation of an OR gate and verify its truth table.

*Introduction:*

The OR gate will produce a 1 at the output when any or all inputs are 1.

*Materials Needed:*

1  Digital logic trainer, *or*

1  7432 quad two-input OR gate DIP IC

1  Light indicator

2  Input switches

Several hookup wires or leads

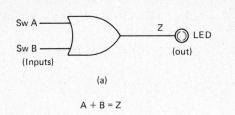

| A | B | Z | LED on or off |
|---|---|---|---|
| 0 | 0 |   |   |
| 0 | 1 |   |   |
| 1 | 0 |   |   |
| 1 | 1 |   |   |

(a)

A + B = Z

(b)

(c)

**Figure 2-12** OR gate: (a) logic symbol; (b) Boolean expression; (c) truth table.

*Procedure:*

1. Connect the OR gate as shown in Figure 2-12a.
2. Set the input switches according to the truth table as shown in the first row.
3. Monitor the output for the proper indication. If the light is off, place a 0 under column Z and write "off" in the LED column. If the light is on, place a 1 under column Z and write "on" in the LED column.
4. Repeat steps 2 and 3 for the remaining rows of the truth table.

*Fill-In Questions:*

1. The OR gate turns on when _____ input is a 1.

2. If one input is a 1 and the other input is a 0, the output for the OR gate will be _____.

3. The Boolean expression for the OR gate is _____.

## EXPERIMENT 3. THE INVERTER

*Objective:*

To show the operation of an inverter.

*Introduction:*

The output of an inverter will always indicate the opposite or complementary condition at its input.

*Materials Needed:*

1  Digital logic trainer, *or*
1  7404 hex inverter DIP IC
1  Light indicator
2  Input switches
Several hookup wires or leads

*Procedure:*

1. Connect the inverter as shown in Figure 2-13a.

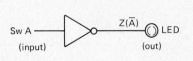

| A | Z | LED on or off |
|---|---|---|
| 0 |   |   |
| 1 |   |   |

(a)

A = Z(Ā)

(b)

(c)

**Figure 2-13** Inverter: (a) logic symbol; (b) Boolean expression; (c) truth table.

2. Set the input switch according to the truth table as shown in the first row.

3. Monitor the output for the proper indication. Place the correct responses in the remaining columns.

4. Activate the switch and perform step 3 again.

*Fill-In Questions:*

1. If the input to an inverter is 0, the output will be _____.

2. If the input to an inverter is 1, the output will be _____.

3. The Boolean expression for an inverter is _____.

## EXPERIMENT 4.   THE NEGATED-INPUT GATE

*Objective:*

To demonstrate the operation of the negated-input gate and verify its truth table.

*Introduction:*

If inverters are placed at some of the inputs to gates, the gates can be turned on with 1s and/or 0s or a combination of both.

*Materials Needed:*

1   Digital logic trainer, *or*

1   7408 quad two-input AND DIP IC

1   7404 hex inverter DIP IC

1   Light indicator

2   Input switches

Several hookup wires or leads

*Procedure:*

1. Connect the circuit shown in Figure 2-14a.

2. Set the input switches according to the truth table as shown in the first row.

3. Monitor the output for the proper indication. Place the correct responses in the remaining columns.

4. Repeat steps 2 and 3 for the remaining rows of the truth table.

**Figure 2-14**   Negated input gate: (a) logic symbol; (b) Boolean expression; (c) truth table.

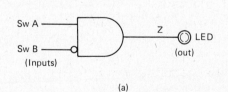

| A | B | Z | LED on or off |
|---|---|---|---|
| 0 | 0 | | |
| 0 | 1 | | |
| 1 | 0 | | |
| 1 | 1 | | |

(a)

(c)

$$A \cdot \overline{B} = Z$$

(b)

*Fill-In Questions:*

1.  The negated-input gate requires a _____ and a _____

    at the inputs to turn it on.

2.  With a 1 to a regular input and a 0 to an inverted input, the output of

    the negated-input gate will be _____.

3.  The Boolean expression for the negated-input gate is _____.

## EXPERIMENT 5.   THE EXCLUSIVE-OR GATE

*Objective:*

To show the operation of the exclusive-OR gate and verify its truth table.

*Introduction:*

The exclusive-OR gate will produce a 1 at the output when one input, and
only one input, is a 1. Refer to Figure 2-5b if this gate must be constructed
from other logic gates.

*Materials Needed:*

*1*  Digital logic trainer, *or*
*1*  7486 quad two-input exclusive-OR DIP IC, *or*
   *1*  7404 hex inverter DIP IC
   *1*  7408 quad two-input AND gate DIP IC
   *1*  7432 quad two-input OR gate DIP IC
*1*  Light indicator
*2*  Input switches
   Several hookup wires or leads

*Procedure:*

1.  Connect the exclusive-OR gate as shown in Figure 2-15a.
2.  Set the input switches according to the truth table as shown in the first
    row.
3.  Monitor the output for the proper indication. Place the correct responses
    in the remaining columns.
4.  Repeat steps 2 and 3 for the remaining rows of the truth table.

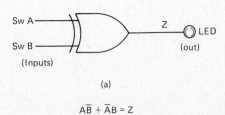

(a)

$$A\overline{B} + \overline{A}B = Z$$

(b)

| A | B | Z | LED on or off |
|---|---|---|---|
| 0 | 0 |   |   |
| 0 | 1 |   |   |
| 1 | 0 |   |   |
| 1 | 1 |   |   |

(c)

**Figure 2-15**  Exclusive-OR gate:
(a) logic symbol; (b) Boolean
expression; (c) truth table.

*Fill-In Questions:*

1.  The exclusive-OR gate turns on when only one _____ is a

    _____.

2.  If all inputs to an exclusive-OR gate are 1, the output will be a _____

    _____.

3.  The Boolean expression for the exclusive-OR gate is _____.

### EXPERIMENT 6.  THE NAND GATE

*Objective:*

To demonstrate the operation of the NAND gate and verify its truth table.

*Introduction:*

The NAND gate will produce a 1 at the output when any input is a 0, or if both inputs are 0.

*Materials Needed:*

1   Digital logic trainer, *or*
1   7400 quad two-input NAND DIP IC
1   Light indicator
2   Input switches
Several hookup wires or leads

*Procedure:*

1.  Connect the NAND gate as shown in Figure 2-16a.
2.  Set the input switches according to the truth table as shown in the first row.
3.  Monitor the output for proper indication. Place the correct responses in the remaining columns.
4.  Repeat steps 2 and 3 for the remaining rows of the truth table.

*Fill-In Questions:*

1.  The NAND gate turns on when any input is a _____.

2.  If one input is a 1 and the other input is a 0, the output for the NAND

    gate will be _____.

3.  The Boolean expression for the NAND gate is _____.

**Figure 2-16**   NAND gate: (a) logic symbol; (b) Boolean expression; (c) truth table.

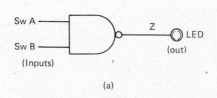

| A | B | Z | LED on or off |
|---|---|---|---|
| 0 | 0 | | |
| 0 | 1 | | |
| 1 | 0 | | |
| 1 | 1 | | |

(a)

$$\overline{A \cdot B} = Z = \overline{A} + \overline{B}$$

(b)

(c)

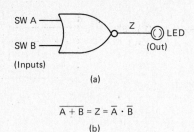

| A | B | Z | LED On or off |
|---|---|---|---|
| 0 | 0 | | |
| 0 | 1 | | |
| 1 | 0 | | |
| 1 | 1 | | |

$$\overline{A + B} = Z = \overline{A} \cdot \overline{B}$$

(b)

(c)

**Figure 2-17** NOR gate: (a) logic symbol; (b) Boolean expression; (c) truth table.

## EXPERIMENT 7.   THE NOR GATE

*Objective:*

To show the operation of the NOR gate and verify its truth table.

*Introduction:*

The NOR gate will produce a 1 at the output when all inputs are 0.

*Materials Needed:*

1   Digital logic trainer, *or*
1   7402 quad two-input NOR DIP IC
1   Light indicator
2   Input switches
    Several hookup wires or leads

*Procedure:*

1.  Connect the NOR gate as shown in Figure 2-17a.
2.  Set the input switches according to the truth table as shown in the first row.
3.  Monitor the output for proper indication. Place the correct responses in the remaining columns.
4.  Repeat steps 2 and 3 for the remaining rows of the truth table.

*Fill-In Questions:*

1.  The NOR gate turns on when _____ inputs are 0.

2.  If one input is a 1 and the other input is a 0, the output for the NOR gate will be _____.

3.  The Boolean expression for the NOR gate is _____.

## EXPERIMENT 8.   THE EXCLUSIVE-NOR GATE

*Objective:*

To demonstrate the operation of the exclusive-NOR gate and verify its truth table.

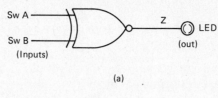

| A | B | Z | LED on or off |
|---|---|---|---|
| 0 | 0 | | |
| 0 | 1 | | |
| 1 | 0 | | |
| 1 | 1 | | |

(a)

(c)

**Figure 2-18** Exclusive-NOR gate: (a) logic symbol; (b) Boolean expression; (c) truth table.

$$AB + \overline{A}\overline{B} = Z$$

(b)

*Introduction:*

The exclusive-NOR gate will produce a 1 at the output when both inputs are the same, either 1 or 0. Refer to Figure 2-8b if this gate must be constructed from other logic gates. Also remember that an exclusive-NOR gate is an exclusive-OR gate followed by an inverter. This gate is a logic comparator.

*Materials Needed:*

1 Digital logic trainer, *or*

1 7486 quad two-input exclusive-OR DIP IC

1 7404 hex inverter DIP IC *or*

   1 7408 quad two-input AND DIP IC

   1 7432 quad two-input OR DIP IC

1 Light indicator

2 Input switches

   Several hookup wires or leads

*Procedure:*

1. Connect the exclusive-NOR gate as shown in Figure 2-18a.
2. Set the input switches according to the truth table as shown in the first row.
3. Monitor the output for the proper indication. Place the correct responses in the remaining columns.
4. Repeat steps 2 and 3 for the remaining rows of the truth table.

*Fill-In Questions:*

1. The exclusive-NOR gate turns on when all inputs are the _____ condition.

2. If all inputs to an exclusive-NOR gate are 1, the output will be a _____ _____.

3. If all inputs to an exclusive-NOR gate are 0, the output will be a _____ _____.

4. The Boolean expression for the exclusive-NOR gate is _____.

**SECTION 2-5**
**INSTANT REVIEW**

An easy way to remember the operation of the various gates is to memorize the condition at the inputs which turns on each gate (with a 1 at the output). Any other condition will not turn it on. The operation of digital logic gates can be summarized as follows:

- The AND gate will have a 1 at its output when all the inputs are 1.
- The OR gate will have a 1 at its output when any or all of its inputs are 1.
- The exclusive-OR gate will have a 1 at its output when only one input is a 1.
- The inverter's output will be the opposite or complement of its input.
- The NAND gate will have a 1 at its output when any of its inputs are 0.
- The NOR gate will have a 1 at its output when all inputs are 0.
- The exclusive-NOR gate will have a 1 at its output when all inputs are the same or equal.

**SECTION 2-6**
**EXERCISES II**

Perform all the exercises in this section before beginning the next section.

1. Considering the fact that a "turned on" logic gate has a 1 at its output, match the gates in column A with the proper input conditions in column B to turn on each gate. Place the description letter designation selected from column B to the left of the numbered gate in column A.

|            | Column A                | Column B                              |
|------------|-------------------------|---------------------------------------|
| _____ 1.   | AND gate                | a. Any input 0                        |
| _____ 2.   | OR gate                 | b. Specific inputs are 0 and 1        |
| _____ 3.   | Inverter                | c. Any or all inputs are 1            |
| _____ 4.   | NAND gate               | d. Output is the same as input        |
| _____ 5.   | NOR gate                | e. All inputs are 1                   |
| _____ 6.   | Exclusive-OR gate       | f. All inputs are 0                   |
| _____ 7.   | Exclusive-NOR gate      | g. Only one input is 1                |
| _____ 8.   | Driver                  | h. Output is the complement of input  |
| _____ 9.   | Negated-input gate      | i. Inputs are the same                |

2. Indicate if the output of each logic gate is 0 or 1 with the inputs shown.

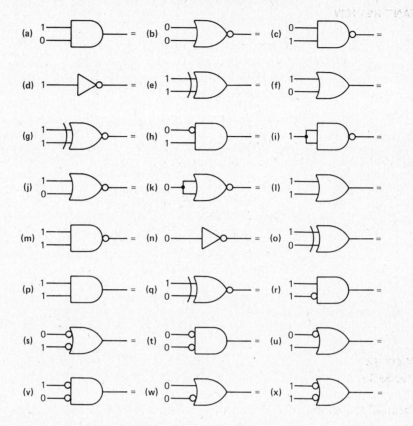

3. Draw the output waveform for each of the logic gates with the inputs shown (observe time relationships $T_1$ through $T_{10}$).

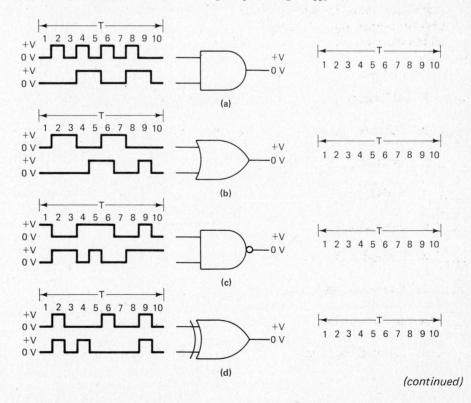

(continued)

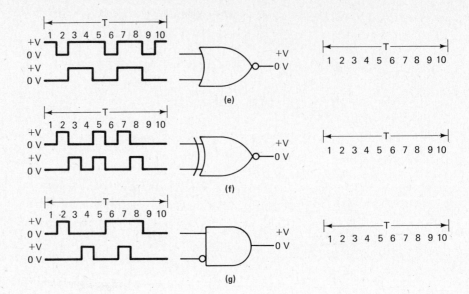

In the following three exercise parts the data at the inputs are represented a little differently, with $1 = +V$ and $0 = 0$ V. The data are grouped into timing charts for easier analysis, with the last line $Z$ (the output) left blank. Draw the output waveform at point $Z$ for each of the logic diagrams on line $Z$ of its respective timing chart according to times $T_0$ through $T_6$.

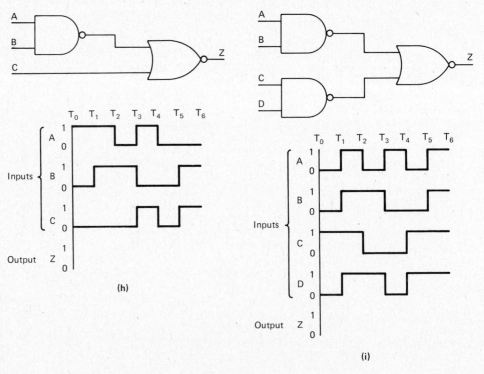

(continued)

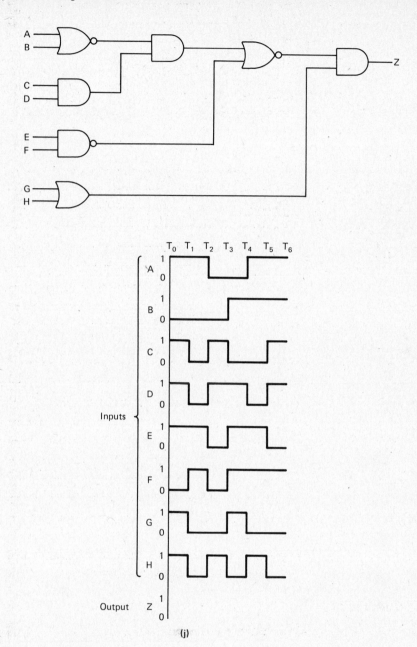

(j)

## SECTION 2-7
## TROUBLESHOOTING APPLICATION: LOGIC GATES

The testing of logic gates can be accomplished by testing each input to the gate for proper action observed at the output. When one input is being tested, it is required that the other inputs be connected to either a fixed 1 or a 0, depending on the function of the logic gate. Figure 2-19 shows the output results for a properly operating gate when one input is tested at a time. Construct each circuit and test the corresponding logic gate (IUT means "input under test").

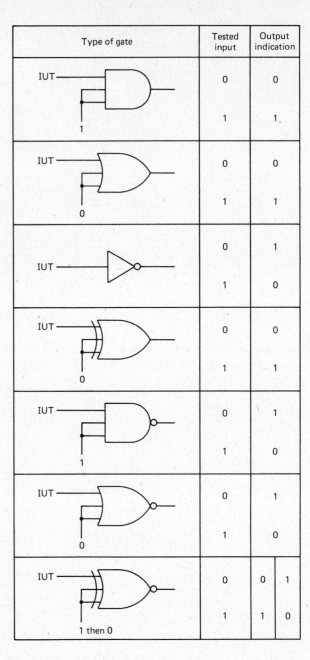

| Type of gate | Tested input | Output indication | |
|---|---|---|---|
| IUT (AND gate) 1 | 0 | 0 | |
| | 1 | 1 | |
| IUT (OR gate) 0 | 0 | 0 | |
| | 1 | 1 | |
| IUT (inverter) | 0 | 1 | |
| | 1 | 0 | |
| IUT (XOR gate) 0 | 0 | 0 | |
| | 1 | 1 | |
| IUT (NAND gate) 1 | 0 | 1 | |
| | 1 | 0 | |
| IUT (NOR gate) 0 | 0 | 1 | |
| | 1 | 0 | |
| IUT (XNOR gate) 1 then 0 | 0 | 0 | 1 |
| | 1 | 1 | 0 |

**Figure 2-19**

## SECTION 2-8
## SELF-CHECKING QUIZZES

### 2-8a  LOGIC GATES: TRUE-FALSE QUIZ

Place a T for true or an F for false to the left of each statement.

_____  1.  Logic gates are also classified as linear circuits.

_____  2.  The exclusive-OR gate will have an output of 1 when its inputs are 1.

_____  3.  When the inputs to a NAND gate are connected together, the output responds as an inverter.

_____ 4. With inputs being 1, 0, and 1 to an AND gate, its output would be 1.

_____ 5. A NOR gate can be connected to serve as an inverter.

_____ 6. Negated inputs to an OR gate cause it to operate the same as a NOR gate.

_____ 7. With inputs 1, 0, and 1 to an OR gate, its output would be 1.

_____ 8. The output of an exclusive-NOR gate will be 1 when its inputs are the same or identical.

_____ 9. Negated inputs to an AND gate cause it to operate the same as a NOR gate.

_____ 10. With inputs 1, 1, 0, and 1 to a NAND gate, its output would be 1.

## 2-8b  LOGIC GATES: MULTIPLE-CHOICE QUIZ

Circle the correct answer for each question.

1. The logic circuit that produces a 1 at the output only when all its inputs are 0 is the:

   a. AND gate        b. OR gate

   c. NAND gate       d. NOR gate

2. The Boolean expression $\overline{A \cdot B}$ represents the:

   a. AND gate        b. OR gate          c. NAND gate

   d. NOR gate        e. none of the above

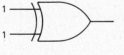     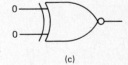

**Figure 2-20**            (a)                              (b)                              (c)

3. The output of the logic gate shown in Figure 2-20a is:

   a. 0     b. 1

4. The output of the logic gate shown in Figure 2-20b is:

   a. 0     b. 1

5. The output of the logic gate shown in Figure 2-20c is:

   a. 0     b. 1

6. With all inputs negated to an OR gate, it has the same operation as the:

   a. AND gate        b. OR gate

   c. NAND gate       d. exclusive-OR gate

7. The Boolean expression $A\overline{B} + \overline{A}B$ represents the:

   a. NAND gate              b. NOR gate

   c. exclusive-OR gate      d. exclusive-NOR gate

8. The correct output of the logic gate shown in Figure 2-21 is:

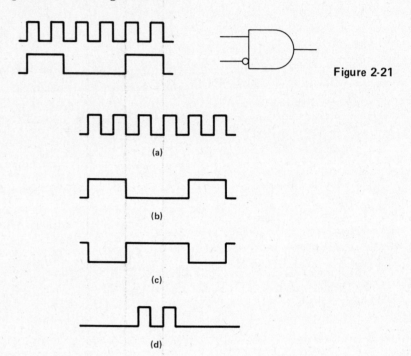

Figure 2-21

(a)

(b)

(c)

(d)

9. The Boolean expression $\overline{A \oplus B}$ represents the:

   a. NAND gate              b. NOR gate

   c. exclusive-OR gate      d. exclusive-NOR gate

10. The Boolean expression $\overline{A + B}$ could be represented by the De Morgan expression:

   a. $A \oplus B$       b. $\overline{AB}$

   c. $\overline{A} \cdot \overline{B}$       d. $\overline{A} + \overline{B}$

## ANSWERS TO EXPERIMENTS AND QUIZZES FOR UNIT 2

*Experiment 1.*

                 (1) all    (2) 0    (3) $AB$

*Experiment 2.*

                 (1) any    (2) 1    (3) $A + B$

*Experiment 3.*

                 (1) 1    (2) 0    (3) $A = \overline{A}$

*Experiment 4.*

                 (1) 0, 1    (2) 1    (3) $A\overline{B}$

*Experiment 5.*

(1) input 1     (2) 0     (3) $A\bar{B} + \bar{A}B$ or $A \oplus B$

*Experiment 6.*

(1) 0     (2) 1     (3) $\overline{AB}$ or $\bar{A} + \bar{B}$

*Experiment 7.*

(1) all     (2) 0     (3) $\overline{A + B}$ or $\bar{A}\bar{B}$

*Experiment 8.*

(1) same     (2) 1     (3) 1     (4) $\overline{A\bar{B} + \bar{A}B}$ or $\overline{A \oplus B}$

*True-False:*

(1) F     (2) F     (3) T     (4) F     (5) T     (6) F     (7) T     (8) T     (9) T
(10) T

*Multiple-Choice:*

(1) d     (2) c     (3) a     (4) a     (5) b     (6) c     (7) c     (8) d     (9) d
(10) c

# Unit 3

---

# Basic Combinational Logic Gates
and Boolean Algebra

**SECTION 3-1**
**IDENTIFICATION, THEORY, AND OPERATION**

**3-1a  INTRODUCTION**

Logic gates are combined into circuits to provide specific functions and/or to control other circuit operations. The manner in which these combinational circuits are designed and constructed uses logic or Boolean algebra (named after its originator, George Boole of England). Knowledge of Boolean algebra aids not only in the design of logic circuits, but also in understanding the operation of circuits for maintenance purposes. This unit introduces you to basic logic circuits, the laws associated with Boolean algebra, and some very helpful hints on troubleshooting logic circuits.

**3-1b  THE AND-TO-OR CIRCUIT**

In the logic diagram shown in Figure 3-1, the output ($AB$) of an AND gate is connected to one input of an OR gate. Input $C$ of the circuit is connected directly to the other input of the OR gate. If inputs $A$ AND $B$ are a 1, OR if $C$ is a 1, the output $AB + C$ will be a 1. An analogy of this logic diagram might be that two small boys ($A$ AND $B$), together can carry a table OR one large boy ($C$) can carry it alone.

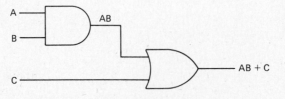

**Figure 3-1**  AND-to-OR logic diagram.

58

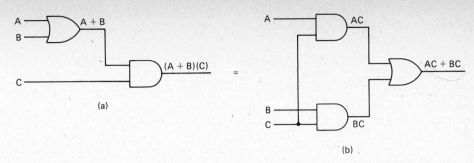

**Figure 3-2** Equivalent logic diagram: (a) OR-to-AND; (b) AND-to-OR.

## 3-1c THE OR-TO-AND CIRCUIT

In the logic diagram shown in Figure 3-2a, the output $(A + B)$ of an OR gate is connected to one input of an AND gate. Input $C$ of the circuit is connected directly to the other input of the AND gate. If input $A$ OR $B$ is a 1 AND $C$ is a 1, the output $A + B(C)$ will be a 1. An analogy of this circuit might be that one of the small boys ($A$ OR $B$) and the larger boy ($C$) together can carry a desk.

If the expression $A + B(C)$ is expanded by logic (Boolean) algebra, which is similar to mathematical algebra, the resulting expression $AC + BC$ forms a different logic diagram, as shown in Figure 3-2b. If inputs $A$ AND $C$ are 1 OR $B$ AND $C$ are 1, the output $AC + BC$ will be a 1. These two logic diagrams are identical; however, the one shown in Figure 3-2a has one fewer logic gate and would probably be preferred.

To prove that these two expressions are identical, a truth table (Table 3-1) can be constructed to show that $A + B(C) = AC + BC$. First, all the possible input conditions are listed in columns 1, 2, and 3. Since there are three inputs, there are eight possible input conditions. Notice how each input condition increases by 1 in binary counting form from 000 until all conditions are 111. The remaining columns show the operations of the expressions, proving their equality. The instruction above each column indicates the operation to be performed on the data in the preceding columns. Column 5, representing output $A + B(C)$, and column 8, representing output $AC + BC$, are identical, proving that the two expressions are equal. This is also an example of the duality of Boolean expressions and the resulting logic diagrams.

### TABLE 3-1

| Input Conditions | | | OR Columns 1 and 2 | AND Columns 3 and 4 | AND Columns 1 and 3 | AND Columns 2 and 3 | OR Columns 6 and 7 |
|---|---|---|---|---|---|---|---|
| *Columns* 1 | 2 | 3 | 4 | 5 | 6 | 7 | 8 |
| A | B | C | A + B | A + B(C) | AC | BC | AC + BC |
| 0 | 0 | 0 | 0 | 0 | 0 | 0 | 0 |
| 0 | 0 | 1 | 0 | 0 | 0 | 0 | 0 |
| 0 | 1 | 0 | 1 | 0 | 0 | 0 | 0 |
| 0 | 1 | 1 | 1 | 1 | 0 | 1 | 1 |
| 1 | 0 | 0 | 1 | 0 | 0 | 0 | 0 |
| 1 | 0 | 1 | 1 | 1 | 1 | 0 | 1 |
| 1 | 1 | 0 | 1 | 0 | 0 | 0 | 0 |
| 1 | 1 | 1 | 1 | 1 | 1 | 1 | 1 |

These columns are identical

59

The order of operations when working with truth tables is as follows:

| Expressions | Procedure |
|---|---|
| 1. $(A + B)(C)$ | OR first, then AND |
| 2. $AC + BC$ | AND first, then OR |
| 3. $\overline{AB}$ | AND first, then invert |
| 4. $\overline{A}\,\overline{B}$ | Invert $A$ and invert $B$ first, then AND |
| 5. $\overline{A + B}$ | OR first, then invert |
| 6. $\overline{A} + \overline{B}$ | Invert $A$ and invert $B$ first, then OR |
| 7. $\overline{\overline{A}\,\overline{B}}$ | Invert $A$ and invert $B$, then AND, then invert |
| 8. $\overline{\overline{A} + \overline{B}}$ | Invert $A$ and invert $B$, then OR, then invert |

### 3-1d LAWS OF BOOLEAN ALGEBRA

The laws of Boolean algebra may be referred to as axioms, postulates, or theorems, which means that they are proven to be true. These laws can be grouped together for easier understanding. The operations of Boolean algebra for commutative, associative, and distributive laws are very similar to the operations of normal algebra.

#### 3-1d.1 Commutative Laws

Law 1.  $AB = BA$

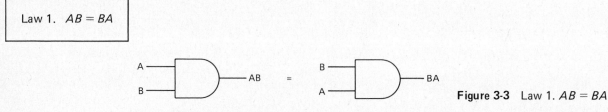

**Figure 3-3**  Law 1. $AB = BA$

This law states that the order in which two variables (or inputs) are ANDed together makes no difference (see Figure 3-3). *Example:* 2 × 3 = 6, 3 × 2 = 6.

Law 2.  $A + B = B + A$

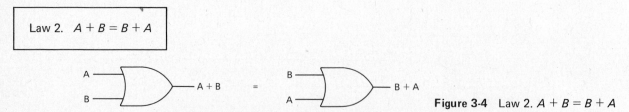

**Figure 3-4**  Law 2. $A + B = B + A$

Similarly, this law states that the order in which two variables are ORed together makes no difference (see Figure 3-4). *Example:* 2 + 3 = 5, 3 + 2 = 5.

#### 3-1d.2 Associative Laws

Law 3.  $(AB)C = A(BC)$

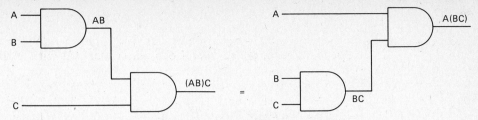

**Figure 3-5** Law 3. $(AB)C = A(BC)$

This law is similar to law 1, but implies three or more variables (see Figure 3-5). *Example:* $(2 \times 3) \times 4 = 6 \times 4 = 24$, $2 \times (3 \times 4) = 2 \times 12 = 24$.

Law 4.   $(A + B) + C = A + (B + C)$

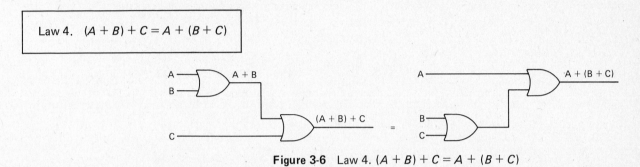

**Figure 3-6** Law 4. $(A + B) + C = A + (B + C)$

This law is similar to law 2, but uses three or more variables (see Figure 3-6). *Example:* $(2 + 3) + 4 = 5 + 4 = 9$, $2 + (3 + 4) = 2 + 7 = 9$.

### 3-1d.3  Distributive Law (Factoring)

Law 5.   $A(B + C) = AB + AC$

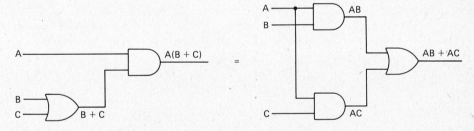

**Figure 3-7** Law 5. $A(B + C) = AB + AC$

This law states that ORing $B$ or $C$ then ANDing with $A$ is equal to ANDing $A$ and $B$, ANDing $A$ and $C$, then ORing $AB$ or $AC$ (see Figure 3-7). It is similar to normal algebra factoring. *Example:* $2(3 + 4) = 2 \times 7 = 14$, $(2 \times 3) + (2 \times 4) = 6 + 8 = 14$.

### 3-1d.4  Complementary Laws

Law 6.   $A \neq \overline{A}, 0 \neq 1, 1 \neq 0$

This law states that $A$ does not equal $\overline{A}$ because 0 does not equal 1 and 1 does not equal 0. However, $\overline{0} = 1$ and $\overline{1} = 0$.

Law 7. $\overline{\overline{A}} = A$

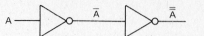

**Figure 3-8**  Law 7. $\overline{\overline{A}} = A$

This law shows that $\overline{\overline{A}} = A$ (see Figure 3-8). If two inverters are connected together and $A$ (the input) = 0, then $\overline{A}$ = 1 and $\overline{\overline{A}}$ (the output) = 0. Similarly, if $A = 1$, then $\overline{A} = 0$ and $\overline{\overline{A}} = 1$.

### 3-1d.5  AND Laws

Law 8.  $A \cdot 0 = 0$

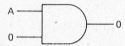

**Figure 3-9**  Law 8. $A \cdot 0 = 0$

This law shows that if one input to an AND function is 0, the output will always be 0 (see Figure 3-9).

Law 9.  $A \cdot 1 = A$

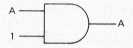

**Figure 3-10**  Law 9. $A \cdot 1 = A$

This law states that if one input to an AND function is a 1, the output depends on the value of $A$ (0 or 1), as in Figure 3-10. It is also known as the *law of intersection*.

Law 10.  $A \cdot A = A$

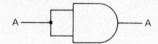

**Figure 3-11**  Law 10. $A \cdot A = A$

This law shows that if both inputs to an AND function are the same, the output will be the same (see Figure 3-11). It is as if only a single piece of wire were used to represent $A$.

Law 11.  $A \cdot \overline{A} = 0$

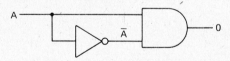

**Figure 3-12**  Law 11. $A \cdot \overline{A} = 0$

This law states that if the two inputs to an AND function are complementary, the output will always be 0 (see Figure 3-12). When $A = 1$, $\overline{A} = 0$, and vice versa.

### 3-1d.6  OR Laws

Law 12.  $A + 0 = A$

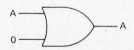

**Figure 3-13**  Law 12. $A + 0 = A$

This law states that if one input to an OR function is 0, the output will depend on the value of $A$ (0 or 1), as in Figure 3-13.

Law 13.  $A + 1 = 1$

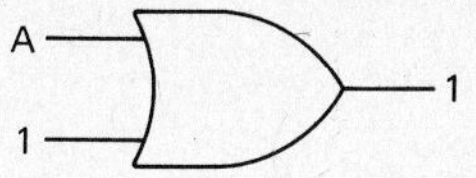

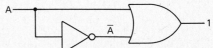

**Figure 3-14**   Law 13. $A + 1 = 1$

This law shows that if one input to an OR function is 1, the output will always be 1 (see Figure 3-14). It is also known as the *law of union*.

Law 14.  $A + A = A$

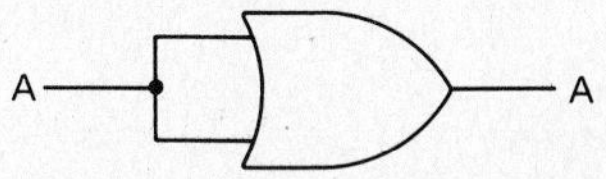

**Figure 3-15**   Law 14. $A + A = A$

This law is similar to law 8 (AND function with both inputs the same) and states that when both inputs to an OR function are the same, the output will also be the same (see Figure 3-15). It, too, is the same as a single piece of wire.

Law 15.  $A + \overline{A} = 1$

**Figure 3-16**   Law 15. $A + \overline{A} = 1$

This law shows that if the two inputs to an OR function are complementary, the output will always be 1 (see Figure 3-16). When $A = 0$, $\overline{A} = 1$, and vice versa.

### 3-1d.7   De Morgan's Laws

Augustus De Morgan (an associate of George Boole) proved that a negated AND function and a negated OR function have an equivalent or dual relationship. This duality exists for all Boolean algebra expressions.

Law 16.  $\overline{AB} = \overline{A} + \overline{B}$

**Figure 3-17**   Law 16. $\overline{AB} = \overline{A} + \overline{B}$

This law states that a negated output AND function is equal to a negated-input OR function (see Figure 3-17). The following expressions are not equal: $\overline{A} \cdot \overline{B} \neq \overline{A} + \overline{B}$; $\overline{AB} \neq \overline{A} + B$.

Law 17.  $\overline{A + B} = \overline{A} \cdot \overline{B}$

**Figure 3-18**   Law 17. $\overline{A + B} = \overline{AB}$

This law states that a negated-output OR function is equal to a negated-input AND function (see Figure 3-18). The following expressions are not equal: $\overline{A} + \overline{B} \neq \overline{AB}$; $\overline{A + B} \neq \overline{AB}$.

**Rules for De Morganizing a Boolean Expression:**

There are two general rules for De Morganizing (finding the equivalent expression) a Boolean expression.

1. Change the operation of the expression, making AND symbols of OR symbols, and vice versa, between the terms. (*Note:* In more complex expressions a group of variables with a solid bar above them may be considered a single term: $\overline{AB}$; $\overline{A + B}$; etc.)
2. Then complement each term.

**Example (a):**

$$\overline{AB} =$$
$$= \overline{\overline{A} + \overline{B}} \quad \text{(Rule 1)}$$
$$= \overline{A} + \overline{B} \quad \text{(Rule 2)}$$

**Example (b):**

$$\overline{A + B} =$$
$$= \overline{\overline{A} \overline{B}} \quad \text{(Rule 1)}$$
$$= \overline{A} \cdot \overline{B} \quad \text{(Rule 2)}$$

**Example (c):**

$$\overline{\overline{AB} + C} =$$
$$= \overline{\overline{AB}} \, (C) \quad \text{(Rule 1)}$$
$$= \overline{\overline{AB}} \, (\overline{C}) \quad \text{(Rule 2)}$$

**Example (d):**

$$\overline{\overline{A + B}(C)} =$$
$$= \overline{\overline{A + B}} + C \quad \text{(Rule 1)}$$
$$= \overline{\overline{A + B}} + \overline{C} \quad \text{(Rule 2)}$$

Since $\overline{\overline{A}} = A$, $\overline{\overline{AB}} = AB$ and the equivalent expression equals $AB\overline{C}$.

Since $\overline{\overline{A}} = A$, $\overline{\overline{A + B}} = A + B$ and the equivalent expression equals $A + B + \overline{C}$.

### 3-1d.8   Reduction Laws

The reduction laws can be used to simplify more complex Boolean expressions and reduce the number of logic gates in a circuit while maintaining the original function. Complex Boolean expressions can be reduced by using the following operations:

1. Expanding (Boolean multiplication)
2. Distribution (multiplying an expression with variables contained in the expression)
3. Factoring
4. Applying the Boolean laws

Groups of similar terms can be treated as single variables in order to apply the basic laws for simplifying a Boolean expression. *Example:* Law 10 = $A \cdot A = A$, similarly, $\overline{A}B \cdot \overline{A}B = \overline{A}B$; law 13 = $A + 1 = 1$, similarly, $\overline{A}B + 1 = 1$. These operations can also be used to prove the reduction laws.

Law 18. $A(A + B) = A$

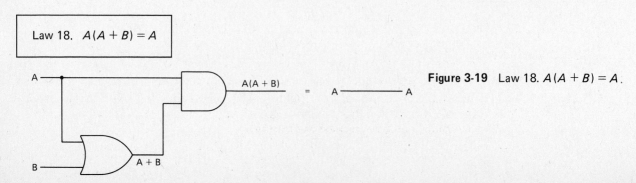

**Figure 3-19** Law 18. $A(A + B) = A$.

Apply $\begin{cases} \text{Expanding:} & A \cdot A + AB = A \\ \text{Law 10 } (A \cdot A = A): & A + AB = A \\ \text{Factoring } A: & A(1 + B) = A \\ \text{Law 13 } (B + 1 = 1): & A(1) = A \\ \text{Law 9 } (A \cdot 1 = A): & A = A \end{cases}$

---

Law 19.  $A + AB = A$

---

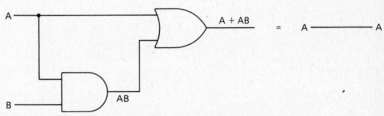

**Figure 3-20**  Law 19. $A + AB = A$

This law was proven in the previous example from law 10 down.

---

Law 20.  $(A + B)(A + C) = A + BC$

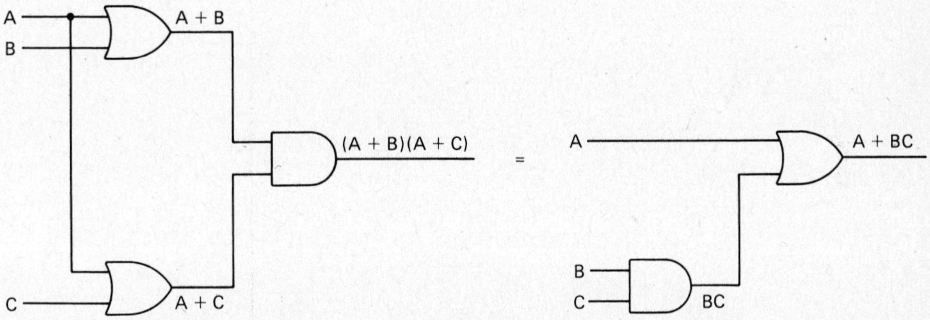

---

**Figure 3-21**  Law 20. $(A + B)(A + C) = A + BC$

Apply $\begin{cases} \text{Expanding:}* & AA + AB + AC + BC = A + BC \\ \text{Law 10 } (A \cdot A = A): & A + AB + AC + BC = A + BC \\ \text{Factoring } A: & A(1 + B + C) + BC = A + BC \\ \text{Law 13 } (1 + B = 1): & A(1 + C) + BC = A + BC \\ \text{Law 13 } (1 + C = 1): & A(1) + BC = A + BC \\ \text{Law 9 } (A \cdot 1 = A): & A + BC = A + BC \end{cases}$

*Hint:

$$\begin{array}{r} A + B \\ A + C \\ \hline A \cdot A + AB + AC + BC \end{array}$$

This may be an easier method of expanding.

Law 21.  $A + \overline{A}B = A + B$

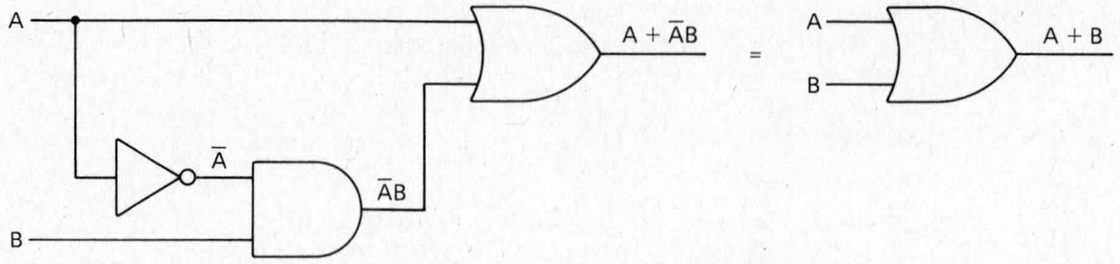

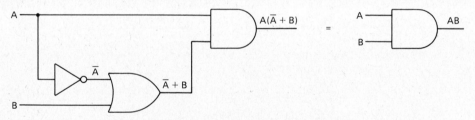

**Figure 3-22**  Law 21. $A + \overline{A}B = A + B$

Apply $\Bigg\{$

| | |
|---|---|
| Distribution $(B + 1)$: | $(B + 1)(A + \overline{A}B) = A + B$ |
| Expanding: | $AB + \overline{A}BB + A + \overline{A}B = A + B$ |
| Law 2, rearrange: | $A + AB + \overline{A}B + \overline{A}BB = A + B$ |
| Law 10 $(B \cdot B = B)$: | $A + AB + \overline{A}B + \overline{A}B = A + B$ |
| Law 14 $(\overline{A}B + \overline{A}B = \overline{A}B)$: | $A + AB + \overline{A}B = A + B$ |
| Factoring $B$: | $A + B(A + \overline{A}) = A + B$ |
| Law 15 $(A + \overline{A} = 1)$: | $A + B(1) = A + B$ |
| Law 9 $(A \cdot 1 = A)$: | $A + B = A + B$ |

Law 22.  $A(\overline{A} + B) = AB$

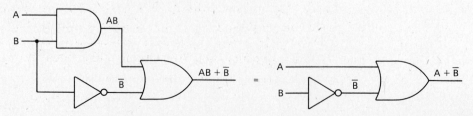

**Figure 3-23**  Law 22. $A(\overline{A} + B) = AB$

Apply $\Bigg\{$

| | |
|---|---|
| Expanding: | $A\overline{A} + AB = AB$ |
| Law 11 $(A\overline{A} = 0)$: | $0 + AB = AB$ |
| Law 12 $(A + 0 = A)$: | $AB = AB$ |

Law 23.  $AB + \overline{B} = A + \overline{B}$

**Figure 3-24**  Law 23. $AB + \overline{B} = A + \overline{B}$

$$\text{Apply} \begin{cases} \text{Distribution } (A + 1): & (AB + \overline{B})(A + 1) = A + \overline{B} \\ \text{Expanding:} & AAB + A\overline{B} + AB + \overline{B} = A + \overline{B} \\ \text{Law 10 } (AA = A): & AB + A\overline{B} + AB + \overline{B} = A + \overline{B} \\ \text{Law 14 } (AB + AB = AB): & AB + A\overline{B} + \overline{B} = A + \overline{B} \\ \text{Factoring } A: & A(B + \overline{B}) + \overline{B} = A + \overline{B} \\ \text{Law 15 } (B + \overline{B} = 1): & A(1) + \overline{B} = A + \overline{B} \\ \text{Law 9 } (A \cdot 1 = A): & A + \overline{B} = A + \overline{B} \end{cases}$$

Law 24.   $(A + B)(\overline{A} + C) = \overline{A}B + AC$

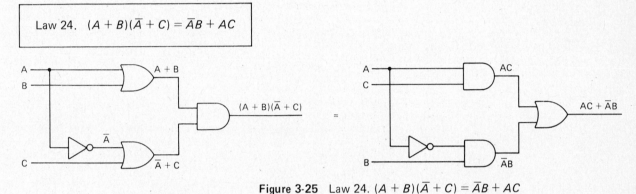

**Figure 3-25**   Law 24. $(A + B)(\overline{A} + C) = \overline{A}B + AC$

$$\text{Apply} \begin{cases} \text{Expanding:} & A\overline{A} + \overline{A}B + AC + BC = \overline{A}B + AC \\ \text{Law 11 } (A \cdot \overline{A} = 0): & \overline{A}B + AC + BC = \overline{A}B + AC \\ \text{Distribution } (\overline{A}B + AC): & \\ & \overline{A}B\overline{A}B + \overline{A}BAC + \overline{A}BBC + AC\overline{A}B + ACAC + ACBC = \overline{A}B + AC \\ \text{Law 10 } (A \cdot A = A): & \\ & \overline{A}B + \overline{A}BAC + \overline{A}BC + AC\overline{A}B + AC + ABC = \overline{A}B + AC \\ \text{Factoring } (\overline{A}B): & \overline{A}B(1 + AC + C + AC) + AC + ABC = \overline{A}B + AC \\ \text{Law 13 } (A + 1): & \overline{A}B + AC + ABC = \overline{A}B + AC \\ \text{Factoring } (AC): & \overline{A}B + AC(1 + B) = \overline{A}B + AC \\ \text{Law 13 } (A + 1): & \overline{A}B + AC = \overline{A}B + AC \end{cases}$$

### 3-1e  SIMPLIFYING BASIC BOOLEAN EXPRESSIONS

Boolean algebra laws are used to reduce or simplify complex expressions to more simple expressions, thereby reducing the number of logic gates used in digital or computer circuits.

#### 3-1e.1  Simplifying 6-Gate-to-2-Gate Logic Diagram

The logic diagram shown in Figure 3-26a will be simplified using Boolean algebra to produce the logic diagram shown in Figure 3-26b. A truth table (Table 3-2) is then constructed to prove that $\overline{A}BC + A\overline{B}C + ABC = C(A + B)$ and the outputs from both circuits will be identical with the input data given.

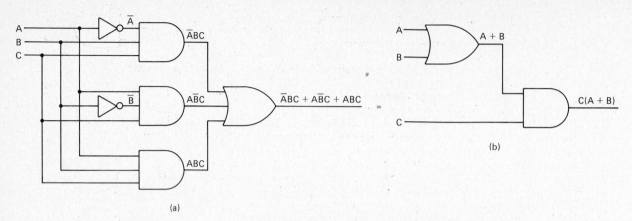

**Figure 3-26**  Six-gate-to-two-gate logic diagram.

**TABLE 3-2**

| A | B | C | $\overline{A}$ | $\overline{A}BC$ | $\overline{B}$ | $A\overline{B}C$ | $ABC$ | $A + B$ | $\overline{A}BC + A\overline{B}C + ABC$ | $C(A + B)$ |
|---|---|---|---|---|---|---|---|---|---|---|
| 0 | 0 | 0 | 1 | 0 | 1 | 0 | 0 | 0 | 0 | 0 |
| 0 | 0 | 1 | 1 | 0 | 1 | 0 | 0 | 0 | 0 | 0 |
| 0 | 1 | 0 | 1 | 0 | 0 | 0 | 0 | 1 | 0 | 0 |
| 0 | 1 | 1 | 1 | 1 | 0 | 0 | 0 | 1 | 1 | 1 |
| 1 | 0 | 0 | 0 | 0 | 1 | 0 | 0 | 1 | 0 | 0 |
| 1 | 0 | 1 | 0 | 0 | 1 | 1 | 0 | 1 | 1 | 1 |
| 1 | 1 | 0 | 0 | 0 | 0 | 0 | 0 | 1 | 0 | 0 |
| 1 | 1 | 1 | 0 | 0 | 0 | 0 | 1 | 1 | 1 | 1 |

Identical

**Simplifying Expressions:**

$$\overline{A}BC + A\overline{B}C + ABC = C(A + B)$$

Factoring AC:     $$\overline{A}BC + AC(\overline{B} + B) = C(A + B)$$

Law 15 $(A + \overline{A} = 1)$:     $$\overline{A}BC + AC(1) = C(A + B)$$

Law 9 $(A \cdot 1 = 1)$:     $$\overline{A}BC + AC = C(A + B)$$

Factoring C:     $$C(\overline{A}B + A) = C(A + B)$$

Law 21 $(A + \overline{A}B = A + B)$:     $$C(B + A) = C(A + B)$$

Law 2 $(A + B = B + A)$:     $$C(A + B) = C(A + B)$$

### 3-1e.2  Simplifying NAND-to-NOR Logic Diagram

The same procedure is used in simplifying this expression (see Figure 3-27 and Table 3-3), but remember to De Morganize the terms under the longest bar first, then the next longest bar, and so on.

**Simplifying Expression:**

$$\overline{\overline{\overline{AB} + C}} = AB\overline{C}$$

De Morgan's rule 1:     $$\overline{\overline{AB}(C)} = AB\overline{C}$$

De Morgan's rule 2:     $$\overline{\overline{AB}}(\overline{C}) = AB\overline{C}$$

Law 7 $(\overline{\overline{A}} = A)$:     $$AB\overline{C} = AB\overline{C}$$

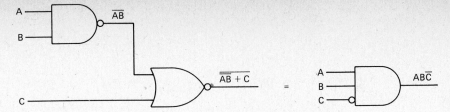

**Figure 3-27**   NAND-to-NOR gate logic diagram.

**TABLE 3-3**

| $A$ | $B$ | $C$ | $AB$ | $\overline{AB}$ | $\overline{AB} + C$ | $\overline{\overline{AB} + C}$ | $\overline{C}$ | $AB\overline{C}$ |
|---|---|---|---|---|---|---|---|---|
| 0 | 0 | 0 | 0 | 1 | 1 | 0 | 1 | 0 |
| 0 | 0 | 1 | 0 | 1 | 1 | 0 | 0 | 0 |
| 0 | 1 | 0 | 0 | 1 | 1 | 0 | 1 | 0 |
| 0 | 1 | 1 | 0 | 1 | 1 | 0 | 0 | 0 |
| 1 | 0 | 0 | 0 | 1 | 1 | 0 | 1 | 0 |
| 1 | 0 | 1 | 0 | 1 | 1 | 0 | 0 | 0 |
| 1 | 1 | 0 | 1 | 0 | 0 | 1 | 1 | 1 |
| 1 | 1 | 1 | 1 | 0 | 1 | 0 | 0 | 0 |

⤷ Identical ⤶

### 3-1e.3   Simplifying NOR-to-NAND Logic Diagram

The same procedure as that used for the preceding problem is used to simplify this expression (see Figure 3-28 and Table 3-4).

**Simplifying Expression:**

$$\overline{\overline{A + B}\,(C)} = A + B + \overline{C}$$

De Morgan's rule 1:   $\overline{\overline{A + B} + C} = A + B + \overline{C}$

De Morgan's rule 2:   $\overline{\overline{A + B}} + \overline{C} = A + B + \overline{C}$

Law 7 $(\overline{\overline{A}} = A)$:   $A + B + \overline{C} = A + B + \overline{C}$

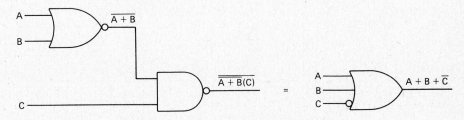

**Figure 3-28**   NOR-to-NAND gate logic diagram.

**TABLE 3-4**

| $A$ | $B$ | $C$ | $A + B$ | $\overline{A + B}$ | $\overline{A + B}(C)$ | $\overline{\overline{A + B}(C)}$ | $\overline{C}$ | $A + B + \overline{C}$ |
|---|---|---|---|---|---|---|---|---|
| 0 | 0 | 0 | 0 | 1 | 0 | 1 | 1 | 1 |
| 0 | 0 | 1 | 0 | 1 | 1 | 0 | 0 | 0 |
| 0 | 1 | 0 | 1 | 0 | 0 | 1 | 1 | 1 |
| 0 | 1 | 1 | 1 | 0 | 0 | 1 | 0 | 1 |
| 1 | 0 | 0 | 1 | 0 | 0 | 1 | 1 | 1 |
| 1 | 0 | 1 | 1 | 0 | 0 | 1 | 0 | 1 |
| 1 | 1 | 0 | 1 | 0 | 0 | 1 | 1 | 1 |
| 1 | 1 | 1 | 1 | 0 | 0 | 1 | 0 | 1 |

⤷ Identical ⤶

**69**

### 3-1f CONSTRUCTING LOGIC DIAGRAMS FROM BOOLEAN EXPRESSIONS

Boolean expressions consist of two forms, which in some cases are interchangeable to accomplish the same logical operation. When the outputs of AND gates are connected to the inputs of OR gates, it is referred to as the *sum-of-products form*. In engineering this is called the *minterm form*. In the other form the outputs of OR gates are connected to the inputs of AND gates; this is called the *product-of sums form*. In engineering this is referred to as the *maxterm form*.

For example, the Boolean expression $AB + CD = Z$ is the sum-of-products or minterm form, and the Boolean expression $(A + B)(C + D) = Z$ is the product-of-sums or maxterm form. Figure 3-25 shows an example of each of these forms.

### SECTION 3-2
### EXERCISES I

Perform all the exercises in this section before beginning the next section.

1. Show the Boolean expression for the output of each logic gate for the following logic diagrams. Develop a truth table for each.

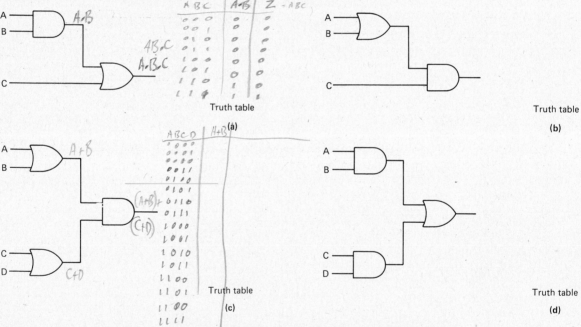

2. Draw the logic diagram for each of the following Boolean expressions. Show the variables at the inputs and the function at the output of each gate.

   a.  $ABC + D = $ output

   b.  $(A + B + C)(D) = $ output

**c.** $AB + \overline{C} =$ output                    **d.** $(A + B)\overline{C} =$ output

3. Indicate the Boolean functions at the output of each logic gate and show the final expression for the following logic diagrams.

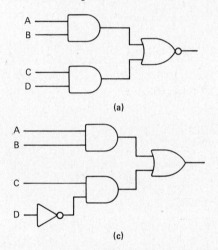

(a)                                                    (b)

(c)                                                    (d)

4. This exercise shows how equivalent logic gates can be connected to produce the same logic function. Four operations are involved:

1. Draw the original logic diagram.
2. Find the equivalent Boolean expression.
3. Construct a truth table to prove the two Boolean expressions' equality.
4. Draw the equivalent logic diagram.

**a.** $AC + ABC = ?$

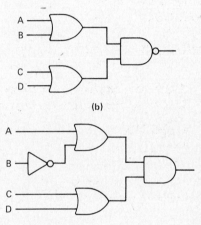

$=$

Original-expression          Equivalent-expression
logic diagram                 logic diagram

Truth table

**b.** $(A + B)(A + C) = ?$

=

Original-expression
logic diagram

Equivalent-expression
logic diagram

Truth table

**c.** $A + \overline{A}B = ?$

=

Original-expression
logic diagram

Equivalent-expression
logic diagram

Truth table

**d.** $A(\overline{A} + B) = ?$

=

Original-expression          Equivalent-expression
logic diagram                    logic diagram

Truth table

5. Perform the following three operations for each problem.

   1. Draw the logic diagram of the original Boolean expression.
   2. Reduce or simplify the expression.
   3. Draw the logic diagram of the equivalent (simplified) expression.

   **a.** $AB + \overline{A}B = ?$

=

*(continued)*

Original-expression  Simplified-expression
  logic diagram   logic diagram

**b.** $AC + \overline{A}BC = ?$

=

Original-expression  Simplified-expression
  logic diagram   logic diagram

**c.** $(A + B)(A + \overline{B}) = ?$

=

Original-expression  Simplified-expression
  logic diagram   logic diagram

**d.** $A\overline{B} + AB + \overline{A}B = ?$

=

Original-expression  Simplified-expression
  logic diagram   logic diagram

e.  $AB + \overline{A}\overline{B} + \overline{A}B + A\overline{B} = ?$

=

Original-expression          Simplified-expression
logic diagram                  logic diagram

6.  For the following forms of Boolean expressions, list the alternate name,
write a simple Boolean expression in that form, and then draw the corresponding logic diagram.

a.  Minterm form: _____
Alternate name

_____
Boolean expression                    Logic diagram

b.  Maxterm form: _____
Alternate form

_____
Boolean expression                    Logic diagram

**SECTION 3-3**
**DEFINITION EXERCISES**

Give a brief description of each of the following terms.

1. Truth table _____

_____

_____

2. Commutative laws (show examples) _____

_____

_____

3. Associative laws (show examples) _____

_____

_____

4. Distributive laws (show examples) _____

_____

_____

5. AND laws (list only expressions) _____

_____

_____

6. OR laws (list only expressions) _____

_____

_____

7. Complementary laws (list only expressions) _____

_____

_____

8. De Morgan's laws (list only expressions) _____

_____

_____

9.  De Morgan's rules _____

_____

_____

10.  Reduction laws (list only expressions) _____

_____

_____

11.  Duality_____

_____

_____

## SECTION 3-4
## EXPERIMENTS

### EXPERIMENT 1.  EQUIVALENT CIRCUITS FOR $AB + AC = A(B + C)$

*Objective:*

To prove visually that the circuits representing the Boolean expressions $AB + AC = A(B + C)$ are identical.

*Introduction:*

Logic switches are wired in parallel to the inputs of the two logic circuits. An LED indicator is placed at the output of each circuit. The table in Figure 3-29b is used to set the inputs $A$, $B$, and $C$. For the inputs, a 0 represents a switch in the down position or on ground, a 1 represents a switch in the up position or $+V$. For the outputs, a 0 represents a LED that is off (not glowing), a 1 represents a LED that is on (glowing). Both LEDs will be on or off for the various conditions set with the input switches.

**Figure 3-29**  Logic circuit for $AB + AC = A(B + C)$: (a) logic diagram; (b) input/output conditions.

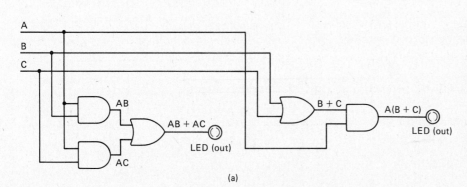

| Inputs | | | Outputs | |
|---|---|---|---|---|
| A | B | C | AB + AC | A(B + C) |
| 0 | 0 | 0 | | |
| 0 | 0 | 1 | | |
| 0 | 1 | 0 | | |
| 0 | 1 | 1 | | |
| 1 | 0 | 0 | | |
| 1 | 0 | 1 | | |
| 1 | 1 | 0 | | |
| 1 | 1 | 1 | | |

(a)                                                                (b)

*Materials Needed:*

    1  Digital logic trainer *or*

    1  7408 quad two-input AND gate IC

    1  7432 quad two-input OR gate IC

    Several hookup wires or leads

*Procedure:*

1. Connect the circuits shown in Figure 3-29a.
2. Connect input logic switches $A$, $B$, and $C$.
3. Set the input switches to the conditions in the first row given in the table of Figure 3-29b.
4. Observe the LEDs for output conditions.
5. In the output section of the table write 0s if the LEDs are off, and 1s if the LEDs are on.
6. Set the input switches for all the input conditions while repeating steps 4 and 5.

*Fill-In Questions:*

1. The Boolean expression $AB + AC$ reads $A$ _____ $B$ _____

    _____ $A$ _____ $C$.

2. The Boolean expression $A(B + C)$ reads $A$ _____ $B$ _____

    _____ $C$.

3. The circuit representing _____ has _____ logic gate

    fewer than the circuit representing $AB + AC$.

## EXPERIMENT 2.  EQUIVALENT CIRCUITS FOR $AC + BC + AD + BD = (A + B)(C + D)$

*Objective:*

To prove visually that the circuit representing the Boolean expressions $AC + BC + AD + BD = (A + B)(C + D)$ are identical.

*Introduction:*

The same as Experiment 1, but with four inputs.

*Materials Needed:*

    1  Digital logic trainer *or*

    2  7408 quad two-input AND gate ICs

    2  7432 quad two-input or gate ICs

        *or*

    1  7432 quad two-input OR gate IC

    1  7425 dual four-input NOR gate IC

    Several hookup wires or leads

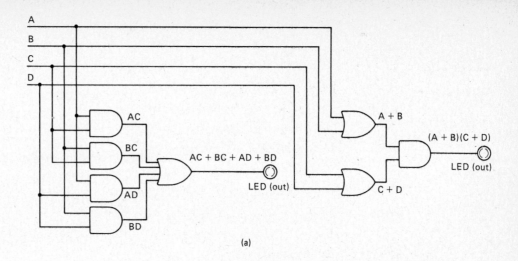

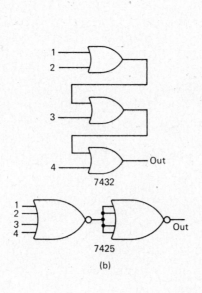

| Inputs | | | | Outputs | | | | | |
|---|---|---|---|---|---|---|---|---|---|
| A | B | C | D | AC + BC + AD + BD | | | | | (A + B)(C + D) |
| 0 | 0 | 0 | 0 | 0 | 0 | 0 | 0 | 0 | 0 |
| 0 | 0 | 0 | 1 | 0 | 0 | 0 | 0 | 0 | 0 |
| 0 | 0 | 1 | 0 | 0 | 0 | 0 | 0 | 0 | 0 |
| 0 | 0 | 1 | 1 | 0 | 0 | 0 | 0 | 0 | 0 |
| 0 | 1 | 0 | 0 | 0 | 0 | 0 | 0 | 1 | 0 |
| 0 | 1 | 0 | 1 | 0 | 0 | 0 | 1 | 1 | 1 |
| 0 | 1 | 1 | 0 | 0 | 1 | 0 | 0 | 1 | 1 |
| 0 | 1 | 1 | 1 | 0 | 1 | 0 | 1 | 1 | 1 |
| 1 | 0 | 0 | 0 | 0 | 0 | 0 | 0 | 1 | 0 |
| 1 | 0 | 0 | 1 | 0 | 0 | 1 | 0 | 1 | 1 |
| 1 | 0 | 1 | 0 | 0 | 0 | 0 | 0 | 1 | 1 |
| 1 | 0 | 1 | 1 | 0 | 0 | 1 | 0 | 1 | 1 |
| 1 | 1 | 0 | 0 | 0 | 0 | 0 | 0 | 1 | 0 |
| 1 | 1 | 0 | 1 | 0 | 0 | 1 | 1 | 1 | 1 |
| 1 | 1 | 1 | 0 | 1 | 1 | 0 | 0 | 1 | 1 |
| 1 | 1 | 1 | 1 | 1 | 1 | 1 | 1 | 1 | 1 |

(c)

**Figure 3-30** Logic circuit for $AC + BC + AD + BD = (A + B)(C + D)$: (a) logic diagram; (b) hints for constructing 4-input OR gate; (c) input/output conditions.

*Procedure:*

1. Connect the circuits shown in Figure 3-30a.
2. Connect inputs switches $A$, $B$, $C$, and $D$.
3. Continue the same procedure as that used in Experiment 1.

*Fill-In Questions:*

1. The Boolean expression $AC + BC + AD + BD$ reads $A$ _____

   $C$ _____ $B$ _____ $C$ _____ $A$ _____

   $D$ _____ $B$ _____ $D$.

2.  The Boolean expression $(A + B)(C + D)$ reads $A$ _____ $B$ _____

_____ $C$ _____ $D$.

3.  Referring to Figure 3-30a, the circuit representing $(A + B)(C + D)$ has

_____ fewer gates than the circuit representing _____.

## EXPERIMENT 3.  EQUIVALENT CIRCUITS FOR $\overline{AB} = \overline{A} + \overline{B}$

*Objective:*

To prove visually that the circuits representing the Boolean expressions $\overline{AB}$ = $\overline{A} + \overline{B}$ are identical.

*Introduction:*

This experiment uses two inputs and proves De Morgan's theorum that a NAND gate function is identical to a negated-input OR function. The procedures used are the same as those used for the previous experiments.

*Materials Needed:*

*1*  Digital logic trainer *or*
*1*  7400 quad two-input NAND gate IC
*1*  7432 quad two-input OR gate IC
*1*  7404 hex inverter IC

         *or*

*1*  7402 quad two-input NOR gate IC
     (see other sections of the book for constructing NOR/NAND inverters)
     Several hookup wires or leads

*Procedure:*

1.  Connect the circuits shown in Figure 3-31a.
2.  Connect input switches $A$ and $B$.
3.  Follow the same procedure as those used in previous experiments.

**Figure 3-31**  Logic circuit for $\overline{AB}$ = $\overline{A} + \overline{B}$: (a) logic diagram; (b) input/output conditions.

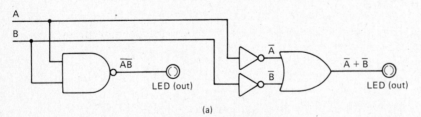

(a)

| Inputs | | Outputs | |
|---|---|---|---|
| A | B | $\overline{AB}$ | $\overline{A} + \overline{B}$ |
| 0 | 0 | | |
| 0 | 1 | | |
| 1 | 0 | | |
| 1 | 1 | | |

(b)

*Fill-In Questions:*

1. The Boolean expression $\overline{AB}$ reads _____ A AND B, or _____

   _____ AND _____ NOT.

2. The Boolean expression $\overline{A} + \overline{B}$ reads _____ A OR _____

   _____ B, or A _____ OR B _____.

3. These two logic gates prove the _____ of Boolean algebra.

## EXPERIMENT 4.   EQUIVALENT CIRCUITS FOR $\overline{A + B} = \overline{A}\overline{B}$

*Objective:*

To prove visually that the circuits representing the Boolean expressions $\overline{A + B} = \overline{A}\overline{B}$ are identical.

*Introduction:*

This experiment uses two inputs and proves De Morgan's theorem that a NOR gate function is identical to a negated-input AND function. The procedures used are the same as those used in previous experiments.

*Materials Needed:*

*1*  Digital logic trainer *or*
*1*  7402 quad two-input NOR gate IC
*1*  7408 quad two-input AND gate IC
*1*  Hex inverter
      *or*
*1*  7400 quad two-input NAND gate IC
      (see other sections of the book for constructing NAND/NOR inverters)
      Several hookup wires or leads

*Procedure:*

1. Connect the circuit shown in Figure 3-32a.
2. Connect input switches A and B.
3. Follow the same procedures as those used in previous experiments.

**Figure 3-32** Logic circuit for $\overline{A + B} = \overline{A}\overline{B}$: (a) logic diagram; (b) input/output conditions.

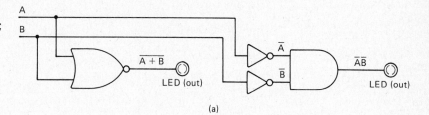

(a)

| Inputs | | Outputs | |
|---|---|---|---|
| A | B | $\overline{A + B}$ | $\overline{A}\overline{B}$ |
| 0 | 0 | | |
| 0 | 1 | | |
| 1 | 0 | | |
| 1 | 1 | | |

(b)

*Fill-In Questions:*

1.  The Boolean expression $\overline{A} + B$ reads _____ *A* or *B*, or _____

    _____ OR _____ NOT.

2.  The Boolean expression $\overline{A}\,\overline{B}$ reads _____ *A* AND _____

    *B*, or *A* _____ AND *B* _____.

3.  These two logic gates prove the _____ of Boolean algebra.

**EXPERIMENT 5.  EQUIVALENT CIRCUITS FOR $\overline{\overline{AB} \cdot \overline{CD}} = AB + CD$**

*Objective:*

To prove visually that the circuits representing the Boolean expressions $\overline{\overline{AB} \cdot \overline{CD}} = AB + CD$ are identical.

**Figure 3-33**   Logic circuit for $\overline{\overline{AB} \cdot \overline{CD}} = AB + CD$: (a) logic diagram; (b) input/output conditions.

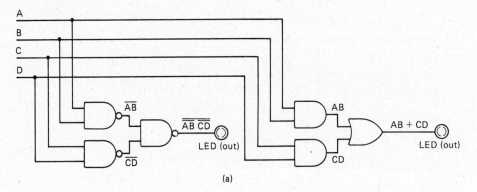

(a)

| Inputs | | | | Outputs | |
|---|---|---|---|---|---|
| A | B | C | D | $\overline{\overline{AB}\ \overline{CD}}$ | AB + CD |
| 0 | 0 | 0 | 0 | | |
| 0 | 0 | 0 | 1 | | |
| 0 | 0 | 1 | 0 | | |
| 0 | 0 | 1 | 1 | | |
| 0 | 1 | 0 | 0 | | |
| 0 | 1 | 0 | 1 | | |
| 0 | 1 | 1 | 0 | | |
| 0 | 1 | 1 | 1 | | |
| 1 | 0 | 0 | 0 | | |
| 1 | 0 | 0 | 1 | | |
| 1 | 0 | 1 | 0 | | |
| 1 | 0 | 1 | 1 | | |
| 1 | 1 | 0 | 0 | | |
| 1 | 1 | 0 | 1 | | |
| 1 | 1 | 1 | 0 | | |
| 1 | 1 | 1 | 1 | | |

(b)

*Introduction:*

This experiment uses four inputs and shows what is known as NAND gate implementation of a Boolean function. The procedures used are the same as those used in previous experiments.

*Materials Needed:*

*1* Digital logic trainer *or*

*1* 7400 quad two-input NAND gate IC

*1* 7408 quad two-input AND gate IC

*1* 7432 quad two-input OR gate IC

Several hookup wires or leads

*Procedure:*

1. Connect the circuit shown in Figure 3-33a.
2. Connect input switches $A$, $B$, $C$, and $D$.
3. Follow the same procedures as those used in previous experiments.

*Fill-In Questions:*

1. The Boolean expression $\overline{\overline{AB}\ \overline{CD}}$ reads NOT _____ $AB$ AND

   _____ $CD$.

2. The Boolean expression $AB + CD$ reads $A$ _____ $B$ _____

   _____ $C$ _____ $D$.

**EXPERIMENT 6.  EQUIVALENT CIRCUITS FOR $\overline{\overline{(A+B)} + \overline{(C+D)}} =$**
**$(A+B)(C+D)$**

*Objective:*

To prove visually that the circuits representing the Boolean expressions $\overline{\overline{(A+B)} + \overline{(C+D)}} = (A+B)(C+D)$ are identical.

*Introduction:*

This experiment uses four inputs and shows NOR gate implementation of a Boolean expression. The procedures used are the same as those used in previous experiments.

*Materials Needed:*

*1* Digital logic trainer *or*

*1* 7402 quad two-input NOR gate IC

*1* 7432 quad two-input OR gate IC

*1* 7408 quad two-input AND gate IC

Several hookup wires or leads

*Procedure:*

1. Connect the circuit shown in Figure 3-34a.
2. Connect input switches $A$, $B$, $C$, and $D$.
3. Follow the same procedures as those used in previous experiments.

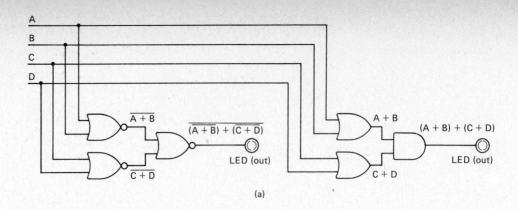

(a)

| Inputs | | | | Outputs | |
|---|---|---|---|---|---|
| A | B | C | D | $\overline{(\overline{A+B})+(\overline{C+D})}$ | $(A+B)+(C+D)$ |
| 0 | 0 | 0 | 0 | | |
| 0 | 0 | 0 | 1 | | |
| 0 | 0 | 1 | 0 | | |
| 0 | 0 | 1 | 1 | | |
| 0 | 1 | 0 | 0 | | |
| 0 | 1 | 0 | 1 | | |
| 0 | 1 | 1 | 0 | | |
| 0 | 1 | 1 | 1 | | |
| 1 | 0 | 0 | 0 | | |
| 1 | 0 | 0 | 1 | | |
| 1 | 0 | 1 | 0 | | |
| 1 | 0 | 1 | 1 | | |
| 1 | 1 | 0 | 0 | | |
| 1 | 1 | 0 | 1 | | |
| 1 | 1 | 1 | 0 | | |
| 1 | 1 | 1 | 1 | | |

(b)

**Figure 3-34** Logic circuit for $\overline{(\overline{A+B})+(\overline{C+D})} = (A+B)(C+D)$: (a) logic diagram; (b) input/output conditions.

*Fill-In Questions:*

1. The Boolean expression $\overline{(\overline{A+B})+(\overline{C+D})}$ reads NOT _____ A

   _____ B OR _____ C _____ D.

2. The Boolean expression $(A+B)(C+D)$ reads A _____ B _____

   _____ C _____ D.

**SECTION 3-5
INSTANT REVIEW**

- Logic gates are combined to produce logic functions that control electronic circuits. Very often a Boolean expression representing a logic function has a dual or equal expression. Similarly, complex logic functions can be reduced or simplified and produce fewer logic gates in the actual circuit. Truth tables can be used to verify the equality of two Boolean expressions.

- The laws of Boolean algebra are grouped below to show their duality. Notice that the dual expression simply has its AND/OR operations reversed, and in some cases 0 becomes 1, and vice versa.

| *Expression* | *Dual* |
|---|---|
| 1. $AB = BA$ | 2. $A + B = B + A$ |
| 3. $(AB)C = A(BC)$ | 4. $(A + B) + C = A + (B + C)$ |
| 5. $A(B + C) = AB + AC$ | 20. $(A + B)(A + C) = A + BC$ |
| 6. $A \neq \overline{A}$ | — |
| 7. $\overline{\overline{A}} = A$ | — |
| 8. $A \cdot 0 = 0$ | 13. $A + 1 = 1$ |
| 9. $A \cdot 1 = A$ | 12. $A + 0 = A$ |
| 10. $A \cdot A = A$ | 14. $A + A = A$ |
| 11. $A \cdot \overline{A} = 0$ | 15. $A + \overline{A} = 1$ |
| 16. $\overline{AB} = \overline{A} + \overline{B}$ | 17. $\overline{A + B} = \overline{A}\overline{B}$ |
| 18. $A(A + B) = A$ | 19. $A + AB = A$ |
| 21. $A + \overline{A}B = A + B$ | 22. $A(\overline{A} + B) = AB$ |
| 23. $AB + \overline{B} = A + \overline{B}$ | |
| 24. $(A + B)(\overline{A} + C) = \overline{A}B + AC$ | |

## SECTION 3-6
## EXERCISES II

Perform all the exercises in this section before beginning the next section.

1. Considering duality, write the dual expression for the following Boolean algebra expressions.

   *Expression*                            *Dual expression*

   a. $A + B = B + A$

   b. $(A + B) + C = A + (B + C)$

   c. $(A + B)(A + C) = A + BC$

   d. $A + 1 = 1$

   e. $A \cdot 1 = A$

   f. $A + A = A$

   g. $A \cdot \overline{A} = 0$

   h. $\overline{A + B} = \overline{A}\overline{B}$

   i. $A(A + B) = A$

   j. $A(\overline{A} + B) = AB$

2. With the inputs given, show the output (either 0 or 1) of each logic gate in the following logic diagrams.

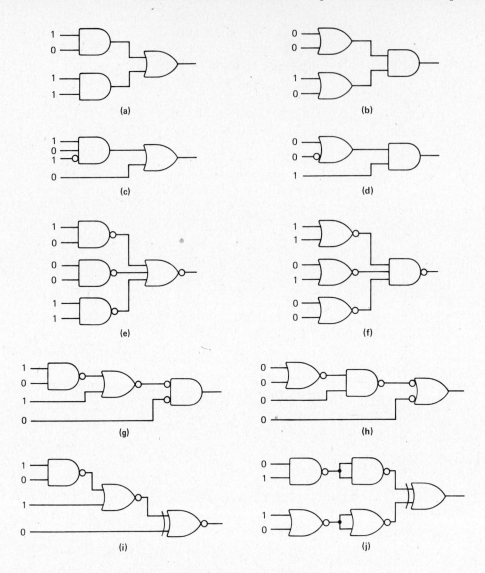

3.  This exercise involves four operations for each of the following problems:

    1.  Write the Boolean expression of the logic diagram.
    2.  Reduce or simplify the expression.
    3.  Construct a truth table to verify the Boolean expressions' equality.
    4.  Draw the logic diagram of the equivalent (simplified) expression.

Example:

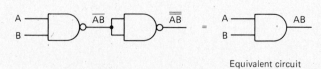

Equivalent circuit

$$\overline{\overline{AB}} = AB$$

| A | B | AB | $\overline{AB}$ | $\overline{\overline{AB}}$ |
|---|---|----|-----|-----|
| 0 | 0 | 0  | 1   | 0   |
| 0 | 1 | 0  | 1   | 0   |
| 1 | 0 | 0  | 1   | 0   |
| 1 | 1 | 1  | 0   | 1   |

Truth table

Truth table

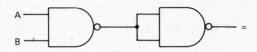

(a)                                    Equivalent logic diagram

Truth table

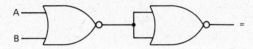

(b)                                    Equivalent logic diagram

Truth table

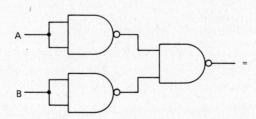

(c)                                    Equivalent logic diagram

*(continued)*

87

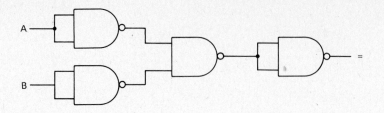

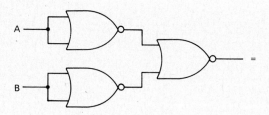

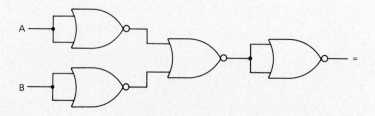

**(f)**    Equivalent logic diagram

Truth table

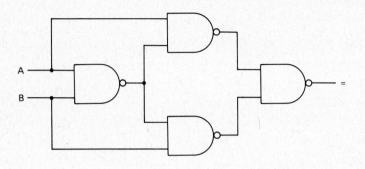

(g)                              Equivalent logic diagram

Truth table

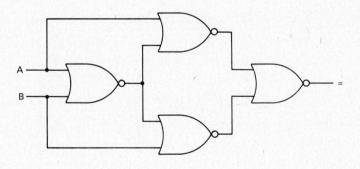

(h)                              Equivalent logic diagram

*(continued)*

Truth table

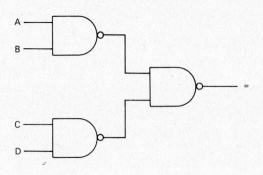

(i)                    Equivalent logic diagram

Truth table

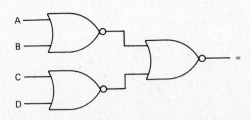

(j)                    Equivalent logic diagram

Truth table

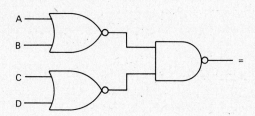

**(k)**          Equivalent logic diagram

Truth table

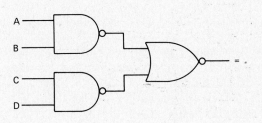

**(l)**          Equivalent logic diagram

**SECTION 3-7**
**TROUBLESHOOTING APPLICATION: COMBINATIONAL LOGIC GATES**

Logic gates can become defective internally and cause problems to other circuits. Locating and correcting these problems can often be a tedious task. Faults at the inputs of a logic gate can cause the output to respond in a definite manner or a variable manner (where the gate may appear normal), depending on the function of the logic gate. In most IC types of logic gates an open input will go high (or become a 1) and the output will respond according to the conditions shown in Figure 3-35.

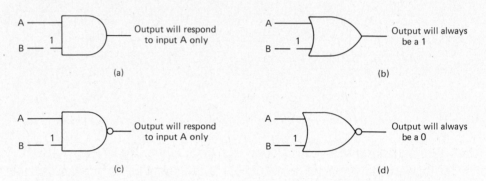

**Figure 3-35**  Logic gates with open input: (a) AND gate; (b) OR gate; (c) NAND gate; (d) NOR gate.

A ground at the input will give a different output condition, as shown in Figure 3-36.

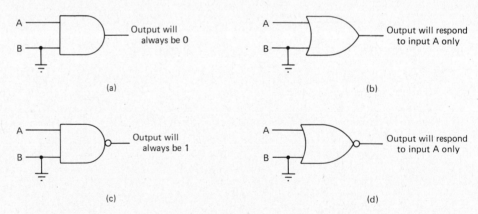

**Figure 3-36**  Logic gates with shorted (grounded) input: (a) AND gate; (b) OR gate; (c) NAND gate; (d) NOR gate.

Logic gates can become shorted together at the inputs, but one way to gain some experience troubleshooting digital circuits is to observe their outputs with solid faults such as opens or shorts at the inputs. Perform the following troubleshooting problems using the step-by-step method given.

1. Construct the circuit.
2. Connect the input logic switches.
3. Set the logic switches for various conditions as discussed in Unit 2, and using a logic probe, examine the various points $X$, $Y$, and $Z$ on the circuit.

4.  Record your response for each point on the circuit in the table. Use the responses: normal, always 0, always 1, and variable (where the output of a gate may appear normal, but another gate connected to one of its inputs is faulty).

Two examples are shown in Figure 3-37. After studying these, record your response for each point in the circuits in Figures 3-38 and 3-39.

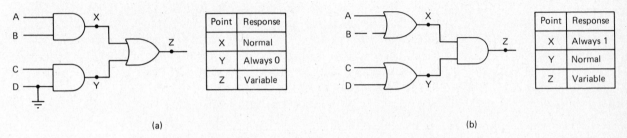

Figure 3-37  Examples of locating faulty logic gates: (a) an input shorted to ground; (b) an input open.

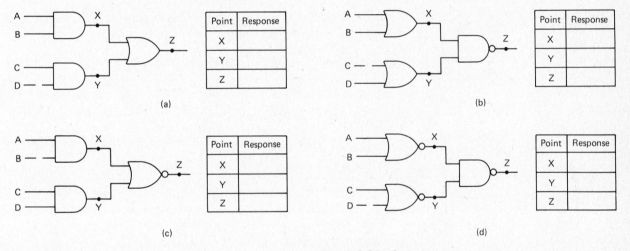

Figure 3-38  Open inputs.

Figure 3-39  Shorted inputs.

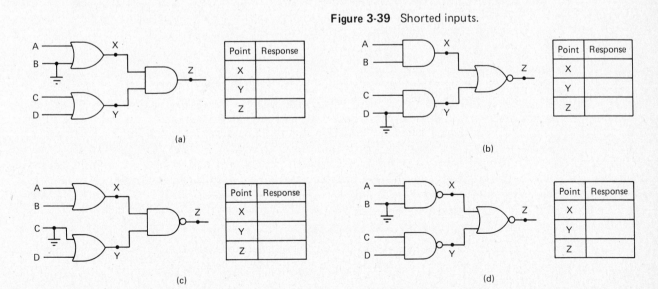

**SECTION 3-8**
**SELF-CHECKING QUIZZES**

### 3-8a  COMBINATIONAL LOGIC GATES: TRUE-FALSE QUIZ

Place a T for true or an F for false to the left of each statement.

_____   1.  Some Boolean expressions can be reduced to simplified equivalent expressions.

_____   2.  The distribution method of Boolean algebra is to AND (Boolean multiplication) an expression with variables contained in the expression to be able to reduce it.

_____   3.  Expanding is referred to as Boolean division.

_____   4.  Boolean factoring is performed in the same manner as normal algebra.

_____   5.  Groups of similar terms may be treated as single variables when reducing Boolean expressions.

_____   6.  The Boolean expression $\overline{A + B}$ is equal to $\overline{A} \cdot \overline{B}$.

_____   7.  A NAND gate with one input grounded will always produce a 0 at the output.

_____   8.  A NOR gate with one input open will always produce a 1 at the output.

_____   9.  To construct a truth table for the Boolean expression $\overline{\overline{A} + \overline{B}}$, you would first invert $A$, invert $B$, then OR, and then invert again.

_____   10.  The Boolean expression $\overline{A} + \overline{B}$ is equal to $\overline{A} \cdot \overline{B}$.

### 3-8b  COMBINATIONAL LOGIC GATES: MULTIPLE-CHOICE QUIZ

Circle the correct answer for each question.

1.  The Boolean expression for Figure 3-40 is:
    a.  $\overline{A} + \overline{B} + \overline{C}$          b.  $\overline{\overline{A \cdot B} + C}$

    c.  $\overline{\overline{A \cdot B} + C}$          d.  $\overline{A} \cdot \overline{B} \cdot \overline{C}$

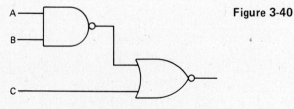

Figure 3-40

2. The final output of the logic circuit in Figure 3-40 would produce a 1 when:

   a. $A = 0$   b. $A = 0$   c. $A = 1$   d. $A = 0$
      $B = 1$      $B = 0$      $B = 1$      $B = 1$
      $C = 0$      $C = 0$      $C = 0$      $C = 1$

3. The Boolean expression $AB + AC + BD + CD$ can be simplified to equal:

   a. $(A + D)(B + C)$          b. $AB + CD$

   c. $(A + B)(A + C)(B + D)(C + D)$   d. $(A + B)(C + D)$

4. Point $W$ of Figure 3-41 would equal:

   a. 0          b. 1

5. Point $X$ of Figure 3-41 would equal:

   a. 0          b. 1

6. Point $Y$ of Figure 3-41 would equal:

   a. 0          b. 1

7. Point $Z$ of Figure 3-41 would equal:

   a. 0          b. 1

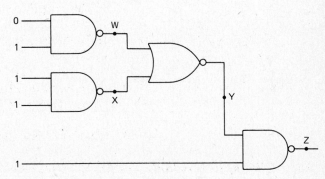

Figure 3-41

8. The Boolean expression $\overline{A}BC + A\overline{B}C + AB\overline{C} + ABC$ can be simplified to equal:

   a. $ABC + A(\overline{B} + \overline{C})$   b. $AB + BC + AC$

   c. $ABC + \overline{A}\overline{B}\overline{C}$   d. $(A + B + C)(\overline{A} + \overline{B} + \overline{C})$

9. The final output of the logic circuit in Figure 3-42 is:

   a. 0          b. 1

Figure 3-42

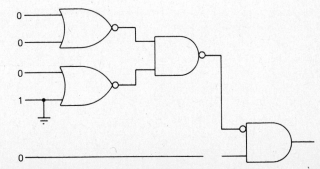

**10.** The Boolean expression $\overline{\overline{ABC} + \overline{A}\overline{B}} + BC$ can be reduced to equal:

a. $\overline{A} \cdot \overline{B} \cdot \overline{C} + A\,B$      b. $A \cdot B$

c. $\overline{A} \cdot \overline{B}$              d. $A \cdot \overline{B} + \overline{A} \cdot B + \overline{C}$

**ANSWERS TO EXPERIMENTS AND QUIZZES FOR UNIT 3**

*Experiment 1.*

(1) AND, OR, AND     (2) AND, OR     (3) $A(B + C)$, one

*Experiment 2.*

(1) AND, OR, AND, OR, AND, OR, AND     (2) OR, AND, OR     (3) two, $AC + BC + AD + BD$

*Experiment 3.*

(1) NOT, $A$, $B$     (2) NOT, NOT, NOT, NOT     (3) duality

*Experiment 4.*

(1) NOT, $A$, $B$     (2) NOT, NOT, NOT, NOT     (3) duality

*Experiment 5.*

(1) NOT, NOT     (2) AND, OR, AND

*Experiment 6.*

(1) NOT, OR, NOT, OR     (2) OR, AND, OR

*True-False:*

(1) T    (2) T    (3) F    (4) T    (5) T    (6) F    (7) F    (8) F    (9) T
(10) F

*Multiple-Choice:*

(1) b    (2) c    (3) a    (4) b    (5) a    (6) a    (7) b    (8) b    (9) b
(10) c

# Unit 4

---

# Clock and Trigger Circuits

## SECTION 4-1
## IDENTIFICATION, THEORY, AND OPERATION

### 4-1a  INTRODUCTION

A circuit called the *clock* is the "heartbeat" in digital systems and computers. It provides the necessary timing pulses for proper operation. A clock is an *astable* or *free-running multivibrator* that switches back and forth between two states, providing a square-edge signal output. It can be considered an electronic oscillator that transforms direct-current voltage into pulsating voltage, thus producing a frequency.

Trigger circuits are used in electronics to provide a clean sharp pulse to other circuits for initiating an action or operation. They can be used to "clean up" or reshape a jagged or distorted pulse, to delay a pulse (or stretch it by making its amplitude of longer duration), and/or to provide a trigger pulse when the amplitude on an input signal reaches a certain point or threshold.

### 4-1b  THE INVERTER CLOCK

A simple clock, shown in Figure 4-1, can be constructed from two inverters, a resistor, and a capacitor. A clock can have two outputs, the $Q$ or true output and its opposite or complementary $\overline{Q}$ output. These outputs can also be referred to as clock phase one ($\phi_1$) and clock phase two ($\phi_2$). When output $Q$ is high or 1, output $\overline{Q}$ is low or 0, and vice versa. The outputs for a normal clock will never be at the same condition or state simultaneously.

A low-value resistor $R_1$ (from 150 to 270 $\Omega$) is used across inverter $I_1$ to bias it near its linear region and therefore must remain fixed. Capacitor $C_1$ is used to provide the necessary feedback for oscillation. It is the main

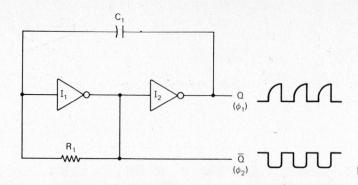

**Figure 4-1**  Inverter clock.

component that determines the frequency of the clock, and therefore is changeable.

The sequence of events for proper circuit action is as follows:

0. Assume that the output of $I_2$ is low. This low is coupled through $C_1$ to the input of $I_1$, causing its output to go high. This high input also at input $I_2$ ensures that the output of $I_2$ will remain low.

1. At this time $C_1$ charges through $R_1$ toward the high at the output of $I_1$.

2. When the voltage at the input to $I_1$ reaches about 1.5 V, its output will go low, forcing the output of $I_2$ to go high. The high output of $I_2$ and the charge on $C_1$ keep the input of $I_1$ high, which keeps the output of $I_1$ low.

3. Capacitor $C_1$ now begins to discharge through $R_1$ toward the low at the output of $I_1$.

4. When $C_1$ discharges to a specific point that establishes a low at the input of $I_1$, the output of $I_1$ goes high, causing the output of $I_2$ to go low, and one cycle is complete. The action will then repeat itself.

The period of one cycle for this clock is approximately $3R_1C_1$, while the frequency of oscillation is the reciprocal of $3R_1C_1$, as given by the formulas

$$\text{period } (p) = 3R_1C_1$$

$$\text{frequency } (f) = \frac{1}{p} = \frac{1}{3R_1C_1}$$

The rounding off of the output pulses is due to the charging and discharging times of $C_1$.

## 4-1c  NAND AND NOR GATE CLOCKS

The output pulses of an inverter clock can be made sharper and more symmetrical with the use of NAND or NOR gates as shown in Figure 4-2. Gates $ND_1$ and $ND_2$, shown in Figure 4-2a, are wired as inverters and produce the same type of outputs as the inverter clock. These outputs are then connected to $ND_3$ and $ND_4$, which form what is called a *bounceless switch* (the bounceless switch is covered in more detail in Unit 5.) When the output of $ND_2$ is 0, it causes the output of $ND_3$ to go to 1, which is applied to one input of $ND_4$. At this time the input of $ND_2$ is 1 and is also applied to the other input of $ND_4$, making its output go to 0. This 0 output is applied

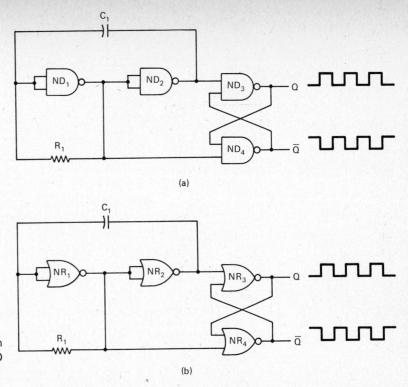

Figure 4-2 Inverter clocks with square-wave outputs: (a) NAND gate clock; (b) NOR gate clock.

to the other input of $ND_3$. When the input of $ND_2$ goes to 0, it causes the output of $ND_4$ to go to 1, which is applied to one input of $ND_3$. At this time the output of $ND_2$ is 1 and is applied to the other input of $ND_3$, which causes its output to go to 0. This action continues as the clock pulses alternate back and forth. The output gates $ND_3$ and $ND_4$ act as buffer circuits for the basic clock and produce symmetrical square-wave pulses.

The NOR gate clock shown in Figure 4-2b works essentially the same as the NAND gate clock. Output frequencies of both types of clocks are found by using the same formula as with the inverter clock.

### 4-1d CRYSTAL-CONTROLLED CLOCK

At high frequencies, usually above 100 kHz, $RC$-type inverter clocks are not stable and therefore are not dependable for critical digital systems. Crystals have long been used in electronics to produce stable oscillators. A crystal can be used with inverters to produce a relatively stable clock, as shown in Figure 4-3.

Resistors $R_1$ and $R_2$ bias the inputs of the inverters to about 1.5 V. With this bias the inverters are within the linear range of operation which is desirable for oscillation when positive or regenerative feedback is present. The crystal (Xtal) provides the feedback for oscillation and the clock will have a frequency dependent on the resonant frequency for which the crystal

Figure 4-3 Crystal-controlled clock.

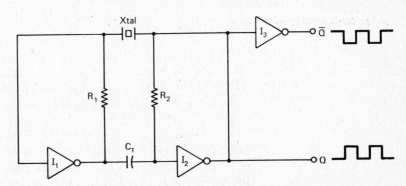

was cut. Capacitor $C_1$ isolates the output of $I_1$ from the bias point at the input of $I_2$ while still passing the pulses needed for oscillation.

A true or $Q$ output is provided at the output of $I_2$ and a complementary or $\overline{Q}$ output is produced by $I_3$.

### 4-1e  THE 555 PRECISION TIMER IC CLOCK

The 555 precision timer IC is a special integrated circuit that can be used for timing, triggering, delaying, pulse reshaping, and clock circuits. A functional diagram is shown in Figure 4-4a. Three 5-k$\Omega$ resistors form a voltage divider from $+V_{CC}$ to ground inside the IC. Two comparators use this voltage divider

**Figure 4-4**  555 precision timer IC: (a) functional diagram; (b) pin identification for mini-DIP; (c) clock circuit.

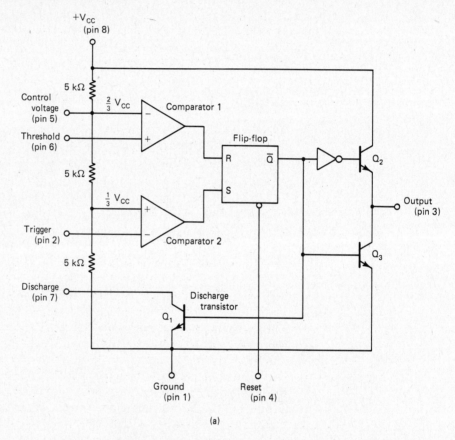

(a)

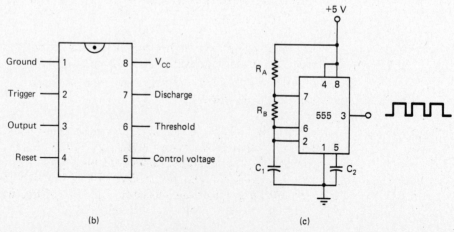

(b)                                                    (c)

to establish voltage reference points at each of one of their inputs. The outputs of the comparators control the operation of an $R$-$S$ flip-flop (flip-flops are covered in more detail in Unit 5.) The $\overline{Q}$ output of the flip-flop goes to an inverter stage, which controls the turning on and off of transistor $Q_2$. This transistor acts as a *source* transistor, connecting the output at pin 3 to the voltage source ($+V_{CC}$). Another output line from the $R$-$S$ flip-flop controls the operation of transistor $Q_3$. This transistor acts as a *sink*, connecting the output pin to ground at pin 1. Only one of these transistors can be on at a single time and is referred to as a *totem-pole output*. For example, if $Q_2$ is on, $Q_3$ is off, and vice versa. The output from the $R$-$S$ flip-flop also controls the operation of $Q_1$, which is used to discharge an external capacitor used for timing sequences at pin 7. Pin 6 (threshold) and pin 2 (trigger) are used for external $RC$ timing components. Pin 4 is used to reset the flip-flop externally and when not used is connected to $+V_{CC}$. Pin 5 can be used for externally controlling the reference voltage on comparator 1. When pin 5 is not used, a 0.01-$\mu$F capacitor is usually connected from it to ground to bypass noise and/or power supply voltage variations to ground, thereby reducing any effects on the reference voltage of comparator 1.

### 4-1e.1  Basic Operation of the 555 Timer IC

1. The reference voltage of comparator 1 is set at $^2/_3 V_{CC}$. When the other input at pin 6 (threshold) is below the value of the reference voltage, the output of the comparator is low or 0, and the flip-flop will not reset. Assuming that the flip-flop is on, there is a low or 0 condition at its output which keeps $Q_1$ and $Q_3$ off, but because of the inverter, places a high or 1 condition at the base of $Q_2$. This transistor turns on and connects the output (pin 3) to $+V_{CC}$ or a high condition.

2. When the voltage at pin 6 reaches $^2/_3 V_{CC}$, a high or 1 condition appears at the output of comparator 1, which resets the flip-flop (turning it off) and the $\overline{Q}$ output goes to a high or 1 condition. At this time, $Q_1$ and $Q_3$ turn on and $Q_2$ turns off. The output is now connected through $Q_3$ to ground or a low condition.

3. The reference voltage of comparator 2 is set at $^1/_3 V_{CC}$. If the other input at pin 2 (trigger) is above the value of the reference voltage, the output of the comparator is low or 0 and the flip-flop will not set (turn on).

4. When the voltage at pin 2 falls to $^1/_3 V_{CC}$, a high or 1 condition appears at the output of comparator 2, which sets (turns on) the flip-flop and its $\overline{Q}$ output goes to a low or 0 condition. Now $Q_1$ and $Q_3$ turn off and $Q_2$ turns on. The output of the 555 is again connected to $+V_{CC}$ or a high and $Q_1$ is off, allowing an external capacitor to charge toward $+V_{CC}$ and repeat the action.

Figure 4-4b shows the pin identification for a commonly used eight-pin mini-dual-in-line-package (mini-DIP) IC.

### 4-1e.2  Basic 555 Clock

A basic 555 timer IC clock is shown in Figure 4-4c. Components $R_A$, $R_B$, and $C_1$ comprise the external $RC$ timing circuit. Capacitor $C_2$ is used to bypass any noise or extraneous voltages to ground.

Capacitor $C_1$ charges toward $+V_{CC}$ through resistors $R_A$ and $R_B$, but only discharges through resistor $R_B$. Therefore, with this basic clock the output pulse is of longer duration when high or a 1 than when low or 0. The frequency of operation can be found by the formula

$$f = \frac{1.49}{(R_A + 2R_B)C_1}$$

### 4-1e.3   555 Duty Cycle

The duty cycle of a device is the amount of time that it is on compared to the time it is off. In the case of the 555 timer clock, it is the ratio of time the output is low ($t_{low}$) to the total period of one cycle ($T$) and can be determined by the formulas

$$t_{low} = 0.693(R_B)C \qquad \text{(the time the output is low)}$$

$$t_{high} = 0.693(R_A + R_B)C \qquad \text{(the time the output is high)}$$

Thus the period ($T$) of one cycle is

$$T = t_{low} + t_{high} = 0.693(R_A + 2R_B)C$$

The duty cycle (DC) is then

$$\text{DC} = \frac{t_{low}}{T}$$

or simply

$$\text{DC} = \frac{R_B}{R_A + 2R_B} \qquad \text{(expressed as a percentage)}$$

Since $C_1$ charges up through $R_A$ and $R_B$, but discharges only through $R_B$, the duty cycle will never be greater than 50%. However, the duty cycle can be controlled from about 5% to greater than 95% by connecting a diode across $R_B$ with the anode at pin 7 and the cathode at pin 6.

### 4-1f   PULSE NOMENCLATURE AND FREQUENCY

In analyzing and troubleshooting digital systems it is important to understand voltage pulse parameters and how to determine frequency. Figure 4-5a shows an ideal pulse as it is generated from an ideal clock. The *leading edge* of a pulse is that part that begins to increase in amplitude from a given reference point. The amount of increase is called the *pulse height* or *amplitude*. The *trailing edge* of a pulse is that part that begins to decrease in amplitude toward a given reference point. The time duration measured between the 50% level in amplitude of the leading edge and the trailing edge of a pulse is called the *pulse width* (PW).

When ideal pulses are applied to circuits, they become distorted because of capacitance and inductive properties of the circuits. This causes time delays and sometimes extraneous pulses which upset the stability of a digital system. The rise and fall time of the leading edge and trailing edge of a pulse become very critical, as shown in Figure 4-5b. The *rise time* ($t_R$) is the time it takes the leading edge of a pulse to increase from 10% to 90% of its maximum amplitude. The *fall time* ($t_F$) is the time it takes the trailing edge of a pulse to decrease from 90% to 10% of its maximum amplitude.

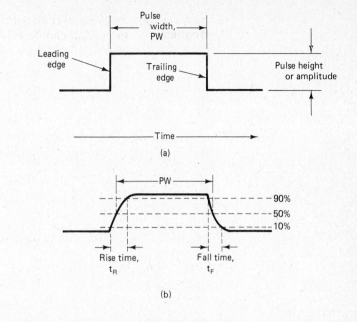

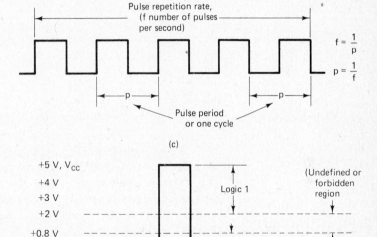

**Figure 4-5**  Voltage pulse nomen-
clature: (a) ideal pulse; (b) actual
pulse showing time delay; (c) pulse
repetition rate (PRR) or frequency;
(d) TTL logic threshold voltage
levels.

The *pulse repetition rate* (PRR) or frequency of a series of pulses can
be determined from the time period of one pulse or cycle, as shown in Figure 4-5c. The time of a single pulse can be measured from the leading edge of
one pulse to the leading edge of the next pulse or from their trailing edges.
The number of pulses per second or frequency can be found by the formula

$$f = \frac{1}{p}$$

where $p$ is the period of one pulse. For example, if $p = 10\ \mu\mathrm{s}$, then

$$f = \frac{1}{p} = \frac{1}{10 \times 10^{-6}\ \mathrm{s}} = 0.1 \times 10^{+6} = 100{,}000 = 100\ \mathrm{kHz}$$

Also, the pulse period of a frequency can very easily be found by the formula

$$p = \frac{1}{f}$$

where $f$ is the frequency. For example, if $f$ = 10 kHz, then

$$p = \frac{1}{f} = \frac{1}{10 \times 10^{+3} \text{ h}} = 0.1 \times 10^{-3} = 0.0001 = 0.1 \text{ ms}$$

Remember, if the pulse period is symmetrical (the high and low states are equal in time), the actual pulse width is half of the pulse period time. For example, if $p$ = 10 $\mu$s, then PW = 5 $\mu$s, or if $p$ = 0.1 ms, then PW = 0.05 ms.

Since solid-state devices are not consistent and actually turn on and off at various voltage levels, a standard must be used to guarantee proper operation of digital circuits. A standard for TTL logic threshold voltages is shown in Figure 4-5d. A *logic 0* is defined as being 0 V up to a threshold of +0.8 V. A *logic 1* is defined as being a threshold of +2.0 V up to +5.0 V. The area from +0.8 V to +2.0 V is called the *undefined* or *forbidden region*. Actually, most digital circuits operate in this region, but they may not be stable (might or might not turn on or off). Therefore, if a digital circuit is going to be guaranteed to operate with a logic 0 at its input, the voltage level must be below +0.8 V. Similarly, if the circuit is supposed to respond to a logic 1 at its input, the voltage level must be above +2.0 V.

### 4-1g   555 TIMER ONE-SHOT TRIGGER CIRCUIT

The *monostable (one-shot) multivibrator* indicated by block diagram form in Figure 4-6a remains in one state or condition until it is triggered. When it is triggered at the input, its output will change state for a time determined by an $RC$ time constant and then return to the initial state. A one-shot trigger circuit can stretch or delay a pulse, reshape a ragged pulse, provide a sharp-leading-edge pulse to trigger other circuits, and be used as a debouncing circuit for mechanical switches. Once a one-shot trigger circuit is in operation, it cannot be triggered by other input pulses until it has finished the timing cycle determined by the $RC$ components.

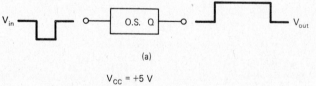

(a)

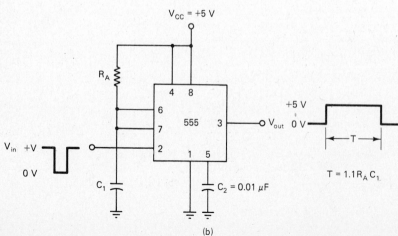

(b)

**Figure 4-6**  555 timer as a one-shot trigger circuit: (a) block diagram of a monostable (one-shot) multivibrator; (b) 555 timer circuit.

The 555 timer IC can be used for a one-shot trigger circuit as shown in Figure 4-6b. Components $R_A$ and $C_1$ determine the duration of time (timing cycle) that the output will be high and the time is determined by the formula

$$T = 1.1 R_A C_1$$

A negative-going pulse must appear at pin 2 to trigger the circuit into its timing cycle, during which time output pin 3 will be high.

### 4-1h  74121 MONOSTABLE MULTIVIBRATOR

The 74121 monostable multivibrator DIP IC functionally contains a flip-flop, an AND gate, and a negated-input OR gate, as shown in Figure 4-7a. The flip-flop has a $Q$ output (pin 6) and a $\overline{Q}$ output (pin 1). Normally, the $Q$ output is low and the $\overline{Q}$ output is high. When the 74121 is triggered, the $Q$ output goes high and the $\overline{Q}$ output goes low during the timing cycle. When the inputs (pins 3 and 4) of $OR_1$ are connected to ground, the 74121 can be triggered by a positive-going pulse to $AND_1$ (pin 5), as shown in Figure 4-7b. If pins 4 and 5 are connected to $+V_{CC}$, a negative-going pulse to pin 3 can trigger the 74121. Similarly, if pins 3 and 5 are connected to $+V_{CC}$, a negative-going pulse to pin 4 will activate the circuit.

**Figure 4-7**  74121 one-shot trigger circuit: (a) typical 74121 monostable multivibrator DIP IC; (b) basic trigger circuit.

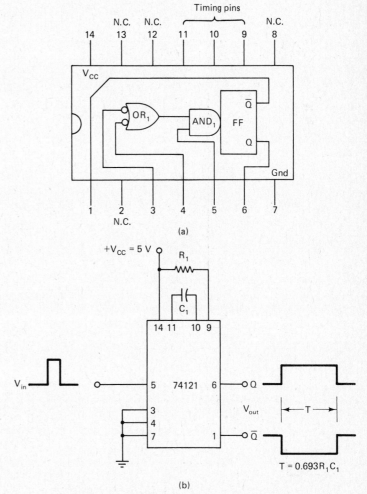

External timing is accomplished by placing a resistor from pin 9 to $+V_{CC}$ and connecting a capacitor across pins 10 and 11. The timing cycle can be calculated by the formula

$$T = 0.693R_1C_1$$

There does exist a monostable multivibrator that can be triggered during its timing cycle to extend the overall time delay. The 74122 DIP IC is a retriggerable monostable multivibrator with a clear input.

## 4-1i  THE SCHMITT TRIGGER CIRCUIT

The *Schmitt trigger circuit* is a circuit whose output changes state when the input is at specific threshold voltage levels. It is recognizable by the hysteresis symbol on its logic function symbol, shown in Figure 4-8a, b, and c.

For a Schmitt trigger inverter, shown in Figure 4-8d, when the input voltage reaches the upper threshold voltage $(+V_t)$ of +1.7 V, the output goes low. When the input voltage falls to the lower threshold voltage $(-V_t)$ of +0.9 V, the output goes high. Figure 4-8e shows how the output swings high and low with the input threshold voltages and constitutes what is called a *hysteresis loop* or *curve* (hence the identifying symbol). The Schmitt trigger circuit is ideal for reshaping pulses and converting ac signals to pulse signals.

**Figure 4-8**  Schmitt trigger: (a) one inverter of the 7414 hex Schmitt-trigger DIP IC; (b) one circuit of the 7413 dual Schmitt-trigger DIP IC; (c) one gate of the quad Schmitt-trigger DIP IC; (d) trigger analysis; (e) hysteresis curve.

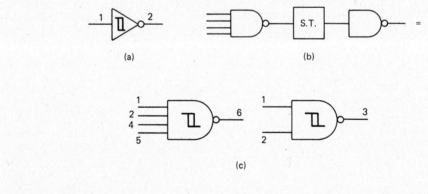

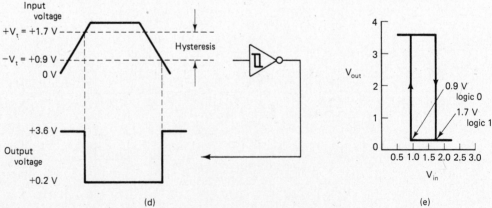

**SECTION 4-2
EXERCISES I**

Perform all the exercises in this section before beginning the next section.

1.  Draw the circuit of a basic inverter clock.

2.  Using the formula $f = 1/3R_1C_1$, find the frequency of an inverter clock
    with the following values.

    a.  $R_1 = 220\ \Omega$               b.  $R_1 = 220\ \Omega$
        $C_1 = 0.1\ \mu F$                    $C_1 = 0.05\ \mu F$

        $f =$ _____              $f =$ _____

    c.  $R_1 = 220\ \Omega$               d.  $R_1 = 180\ \Omega$
        $C_1 = 0.2\ \mu F$                    $C_1 = 0.1\ \mu F$

        $f =$ _____              $f =$ _____

3.  Draw the circuit of a basic NAND gate, inverter-type clock.

4.  Draw the circuit of a basic NOR gate, inverter-type clock.

5. Draw the circuit of a basic crystal-controlled clock using inverters.

6. Draw the top view of a 555 mini-DIP IC and show the function of each pin.

7. Draw the circuit for a 555 timer IC clock.

8. Using the formulas

$$f = \frac{1.49}{(R_A + 2R_B)C} \qquad t_{low} = 0.693(R_B)C \qquad t_{high} = 0.693(R_A + R_B)C$$

$$T = t_{low} + t_{high} \text{ or } 0.693(R_A + 2R_B)C$$

and

$$DC = \frac{R_B}{R_A + 2R_B}$$

solve the following problems with the data given.

    **a.**  $R_A = 1\ \text{k}\Omega$             **b.**  $R_A = 4.7\ \text{k}\Omega$
        $R_B = 1\ \text{k}\Omega$                 $R_B = 2.2\ \text{k}\Omega$
        $C = 0.01\ \mu\text{F}$              $C = 0.05\ \mu\text{F}$

        $f =$ _____           $f =$ _____

        $t_{\text{low}} =$ _____       $t_{\text{low}} =$ _____

        $t_{\text{high}} =$ _____       $t_{\text{high}} =$ _____

        $T =$ _____          $T =$ _____

        $\text{DC} =$ _____        $\text{DC} =$ _____

9. Find the PRR or frequency of the pulse trains with the following periods given. Use the formula $f = 1/p$.

    **a.**  $p = 0.2\ \text{ms}$            **b.**  $p = 20\ \mu\text{s}$

        $f =$ _____           $f =$ _____

    **c.**  $p = 1\ \mu\text{s}$             **d.**  $p = 33\ \mu\text{s}$

        $f =$ _____           $f =$ _____

10. Find the period (or time of one cycle) for the frequencies given. Use the formula $p = 1/f$.

    **a.**  $f = 15\ \text{kHz}$          **b.**  $f = 250\ \text{kHz}$

        $p =$ _____           $p =$ _____

    **c.**  $f = 1\ \text{MHz}$           **d.**  $f = 2.2\ \text{MHz}$

        $p =$ _____           $p =$ _____

11. Draw a voltage pulse and show the TTL threshold voltage levels.

12. Draw a basic 555 timer one-shot trigger circuit.

13. Draw a basic 74121 one-shot trigger circuit.

14. Draw a Schmitt trigger inverter and show the relationship of the input voltage to the output voltage.

**SECTION 4-3**
**DEFINITION EXERCISES**

Give a brief description of each of the following terms.

1. Clock _____

_____

_____

2. Astable multivibrator _____

_____

_____

3. Free-running multivibrator _____

_____

_____

4. Trigger circuit _____

_____

_____

5. Monostable multivibrator _____

_____

_____

6. One-shot multivibrator _____

_____

_____

7. Single-shot multivibrator _____

_____

_____

8. Regenerative feedback _____

_____

_____

9. $Q$ output _____

_____

_____

10. $\overline{Q}$ output _____

_____

_____

11. Source transistor _____

_____

_____

12. Sink transistor _____

_____

_____

13. Duty cycle _____

_____

_____

14. Schmitt trigger circuit _____

_____

_____

15. Hysteresis _____

_____

_____

16. Voltage threshold _____

_____

_____

17. Leading edge _____

_____

_____

18. Trailing edge _____

_____

_____

19. Pulse width _____

_____

_____

20. Rise time _____

_____

_____

21.  Fall time _____
_____
_____

22.  Amplitude _____
_____
_____

23.  Pulse repetition rate (PRR) _____
_____
_____

24.  Frequency _____
_____
_____

25.  Pulse period _____
_____
_____

26.  TTL voltage levels for a logic 1 _____
_____
_____

27.  TTL voltage levels for a logic 0 _____
_____
_____

28.  Totem-pole output _____
_____
_____

## SECTION 4-4
## EXPERIMENTS

The experiments in this section can be performed on digital logic trainers with the aid of clip leads to connect the required passive components, or they can be wired directly to the ICs using a standard breadboard.

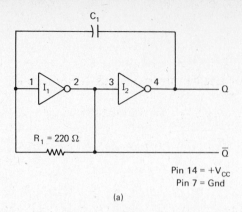

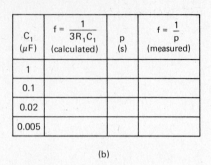

| $C_1$ ($\mu$F) | $f = \dfrac{1}{3R_1C_1}$ (calculated) | p (s) | $f = \dfrac{1}{p}$ (measured) |
|---|---|---|---|
| 1 | | | |
| 0.1 | | | |
| 0.02 | | | |
| 0.005 | | | |

Pin 14 = +$V_{CC}$
Pin 7 = Gnd

(a)  (b)

**Figure 4-9** Inverter clock: (a) circuit; (b) data table.

## EXPERIMENT 1.  INVERTER CLOCK

*Objective:*

To show how a simple clock can be constructed with inverters, using a bias resistor and a feedback capacitor.

*Introduction:*

Since $R_1$ must be about 200 $\Omega$, the capacitor $C_1$ is the determining factor for the frequency of the inverter clock. The $Q$ output is the true output and the $\overline{Q}$ output is the complementary output.

*Materials Needed:*

1   Digital logic trainer *or*
1   7404 hex inverter DIP IC
1   220-$\Omega$ resistor at 0.5 W ($R_1$)
1   1-$\mu$F capacitor at 10 WV dc
1   0.1-$\mu$F capacitor at 10 WV dc
1   0.02-$\mu$F capacitor at 10 WV dc   $\Big\}$ ($C_1$)
1   0.005-$\mu$F capacitor at 10 WV dc
1   Dual-trace oscilloscope
    Several hookup wires or leads

*Procedure:*

1.  Construct the circuit as shown in Figure 4-9a. Use the value of $C_1$ as given in the first column of the first row in the data table of Figure 4-9a.
2.  Place channel 1 of the oscilloscope at the $Q$ output and adjust for a pulse width of about 1 cm.
3.  Place channel 2 of the oscilloscope at the $\overline{Q}$ output. Notice the difference of the outputs.
4.  Calculate $f$ and list it in the second column of the data table.
5.  Measure the period ($p$) of one cycle and list in the third column of the data table.
6.  Calculate the frequency of the clock with the formula $f = 1/p$ and list the answer in the fourth column of the data table. A large difference between the calculated $f$ and the measured $f$ is probably due to the tolerances of $R_1$ and $C_1$.
7.  Change $C_1$ and repeat steps 4 through 6 for the remaining values of $C_1$ given in the data table.

*Fill-In Questions:*

1. The output of a basic inverter clock is not a good _____ wave, because of the charging and discharging of $C_1$.

2. When the value of $C_1$ increases in the inverter clock, the frequency _____.

3. To increase the frequency of the inverter clock, the value of $C_1$ must be _____.


**EXPERIMENT 2.  NAND/NOR CLOCKS**

*Objective:*

To demonstrate how NAND gates and NOR gates can be used to construct basic inverter-type clocks.

*Introduction:*

Two gates of an IC package are wired as the basic inverter clock. The two remaining gates in the package are then used in a cross-connected output which produces good square-wave pulses.

*Materials Needed:*

*1* Digital logic trainer *or*

*1* 7400 quad two-input NAND gate DIP IC

*1* 7402 quad two-input NOR gate DIP IC

*1* 220-Ω resistor at 0.5 W ($R_1$)

*1* 1-$\mu$F capacitor at 10 WV dc ($C_1$)

*1* 0.1-$\mu$F capacitor at 10 WV dc ($C_1$)

*1* Dual-trace oscilloscope

  Several hookup wires or leads

*Procedure:*

1. Construct the circuit shown in Figure 4-10a.
2. Connect channel 1 of the oscilloscope to the Q output and channel 2 to the $\overline{Q}$ output. Observe that the voltage waveforms are better square waves than with the basic inverter clock.

**Figure 4-10**  NAND/NOR clocks: (a) NAND gate clock; (b) NOR gate clock.

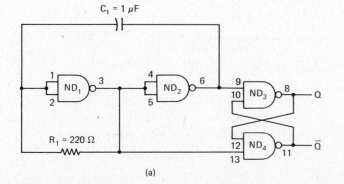
(a)

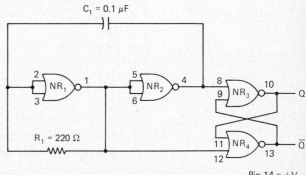
(b)

Pin 14 = +$V_{CC}$
Pin 7 = Gnd

3. Calculate the frequency of the clock from the formula $f = 1/p$ and list

here: $f =$ _____ hertz.

4. Disassemble this circuit and construct the circuit shown in Figure 4-10b.

5. Connect the oscilloscope to the outputs, $Q$ and $\overline{Q}$, and observe the voltage waveforms.

6. Calculate the frequency of this clock from the formula $f = 1/p$ and list

here: $f =$ _____ hertz.

*Fill-In Questions:*

1. Connecting together the inputs of a NAND gate or a NOR gate produces

an _____ function.

2. The cross-connected output gates of a NAND/NOR gate clock produces

good output _____ waves.

3. The formula for approximating the frequency of the NAND/NOR gate

clocks is _____.

**EXPERIMENT 3.   CRYSTAL-CONTROLLED CLOCK**

*Objective:*

To provide experience in using a crystal-controlled clock.

*Introduction:*

Resistors $R_1$ and $R_2$ provide the dc bias for circuit operation. Capacitor $C_1$ isolates the output of $I_1$ from the input of $I_2$. The crystal provides the feedback for oscillation. Resistor $R_3$ may be used to develop a more-square-wave pulse at the output.

*Materials Needed:*

*1*  Digital logic trainer *or*
*1*  7404 hex inverter DIP IC
*2*  1 kΩ resistors at 0.5 W ($R_1$ and $R_2$)
*1*  1- to 10-kΩ resistor at 0.5 W ($R_3$)
*1*  Crystal, within the range 100 kHz to 1 MHz
*1*  Dual-trace oscilloscope
    Several hookup wires or leads

*Procedure:*

1. Construct the circuit shown in Figure 4-11.

2. Connect the dual-trace oscilloscope to the $Q$ and $\overline{Q}$ outputs. Observe the output-voltage waveforms.

3. List the frequency of the crystal in the appropriate place in Figure 4-11.

4. Using the formula $f = 1/p$, calculate the frequency from the waveform shown on the oscilloscope and list it in the appropriate place in Figure

4-11. Does this frequency match that of the crystal? _____.

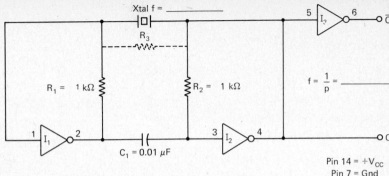

**Figure 4-11** Crystal-controlled clock.

*Fill-In Questions:*

1. The frequency of a crystal-controlled clock is determined by the _____ _____ .

2. This type of clock is very _____ .

3. The output of a basic crystal-controlled clock is _____ the best square wave.

## EXPERIMENT 4.   555 TIMER IC CLOCK

*Objective:*

To demonstrate how the 555 timer IC can be used as a clock and be able to calculate its output frequency and duty cycle.

*Introduction:*

With only two resistors and a capacitor, the 555 timer IC can produce a highly stable clock, as shown in Figure 4-12a. The output frequency and the duty cycle can be calculated using the values of $R_A$, $R_B$, and $C_1$. If a diode is placed across $R_B$, the output voltage waveform becomes symmetrical and has a duty cycle of 50%. Capacitor $C_1$ now charges up through the diode and $R_A$ and discharges through $R_B$.

*Materials Needed:*

1   +5-V power supply
1   Oscilloscope
1   555 timer IC
1   2.2-kΩ resistor at 0.5 W ($R_B$)
2   10-kΩ resistors at 0.5 W ($R_A$, $R_B$)
1   22-kΩ resistor at 0.5 W ($R_A$, $R_B$)
1   0.01-μF capacitor at 25 WV dc ($C_2$)
1   0.05-μF capacitor at 25 WV dc ($C_1$)
1   0.1-μF capacitor at 25 WV dc ($C_1$)
1   0.2-μF capacitor at 25 WV dc ($C_1$)
1   IN4002 diode or equivalent
1   Breadboard for constructing circuit
    Several hookup wires or leads

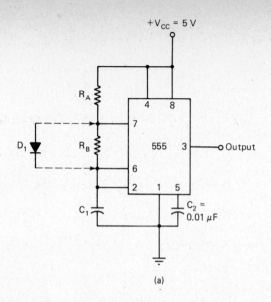

+V_CC = 5 V

(a)

| $R_A$ (Ω) | $R_B$ (Ω) | $C_1$ (μF) | $f_{out} = \dfrac{1.49}{(R_A + 2R_B)C}$ calculated (Hz) | $f_{out}$ measured (Hz) | Output waveform | | $dc = \dfrac{R_B}{R_A + 2R_B}$ |
|---|---|---|---|---|---|---|---|
| 22 k | 10 k | 0.1 | | 0 V | | $\overline{V\ \text{p-p}}$ | |
| 22 k | 10 k | 0.05 | | 0 V | | $\overline{V\ \text{p-p}}$ | |
| 22 k | 10 k | 0.2 | | 0 V | | $\overline{V\ \text{p-p}}$ | |
| 22 k | 2.2 k | 0.1 | | 0 V | | $\overline{V\ \text{p-p}}$ | |
| 10 k | 10 k | 0.1 | | 0 V | | $\overline{V\ \text{p-p}}$ | |

(b)

Figure 4-12  555 timer IC clock: (a) clock circuit; (b) data table. (From F. Hughes, *Basic Electronics: Theory and Experimentation,* Prentice-Hall, Englewood Cliffs, N.J., © 1984, Fig. 12-10, p. 278. Reprinted with permission.)

*Procedure:*

1.  Construct the circuit shown in Figure 4-12a. Use the values of $R_A$, $R_B$, and $C_1$ as given in the first row of the data table of Figure 4-12b. Do not connect diode $D_1$ at this time.

2.  Calculate $f_{out}$ and record in the data table in the first row.

3.  Using the oscilloscope at the output (pin 3), measure $f_{out}$ and record in the data table in the proper place.

4.  Draw the output voltage waveform in the data table and indicate peak-to-peak voltage.

5.  Using the values of $R_A$ and $R_B$, calculate the duty cycle and record it in the data table at the place indicated.

6.  Repeat steps 2 through 5 for the remaining values of $R_A$, $R_B$, and $C_1$ given in the data table.

7.  Using $R_A = 10\ \text{k}\Omega$, $R_B = 10\ \text{k}\Omega$, and $C_1 = 0.1\ \mu\text{F}$, add diode $D_1$ across $R_B$ as shown in Figure 4-12a. Since $R_A = R_B$ and the diode is in place, the output frequency can be determined by the formula

$$f_{out} = \frac{1.49}{2R_A C_1}$$

Calculate the $f_{out}$ and record here. _____

8.  Measure $f_{out}$ with the oscilloscope and notice that the duty cycle is 50%.

*Fill-In Questions:*

1. The $f_{out}$ of a 555 timer clock is determined by _____,

   _____, and _____.

2. If the resistance or capacitance is increased, $f_{out}$ _____.

3. If the resistance or capacitance is decreased, $f_{out}$ _____.

4. Without the use of the diode, the output voltage waveform will be

   _____ and the duty cycle will be _____ than 50%.

5. With the use of the diode across $R_B$ and $R_A = R_B$, the output voltage

   waveform will be _____ and the duty cycle will be _____

   _____%.

6. With the use of a diode across $R_B$ and the proper values of $R_A$ and $R_B$,

   the duty cycle can range from _____ to _____%.

## EXPERIMENT 5.  TWO-TONE ALARM CIRCUIT

*Objective:*

To demonstrate a circuit application using inverter clocks and to show the controlled gating action of logic gates.

*Introduction:*

The heart of this circuit consists of three clocks made from a hex inverter IC. One clock has a high tone frequency of about 1.5 kHz, the second has a low tone frequency at about 750 Hz, and the third operates as a switching circuit with a frequency of about 1.5 Hz. The high tone output is connected to $AND_1$ together with one output of the switching circuit. The low tone output is connected to $AND_2$ together with the other output of the switching circuit. These two AND gates feed $OR_1$, which in turn is connected to transistor $Q_1$ to produce sound at the speaker. When the output of $I_4$ is high (output of $I_3$ is low), the high tone is fed through $AND_1$ and $OR_1$ to the speaker circuit. When the output of $I_3$ is high (output of $I_4$ is low), the low tone is fed through $AND_2$ and $OR_1$ to the speaker circuit. As a result, there is a varying two-tone sound at the speaker.

*Materials Needed:*

- *1*  Digital logic trainer *or*
- *1*  7404 hex inverter DIP IC ($I_1$ to $I_6$)
- \* { *1*  7408 quad two-input AND gate DIP IC ($AND_1$ and $AND_2$)\*
- *1*  7432 quad two-input OR gate DIP IC ($OR_1$)\*
- *3*  220-$\Omega$ resistors at 0.5 W ($R_1$, $R_2$, and $R_3$)
- *1*  10-k$\Omega$ resistor at 0.5 W ($R_4$)

\*A 7451 dual two-wide two-input AND-OR-Inverter gate DIP IC may be substituted for these two ICs.

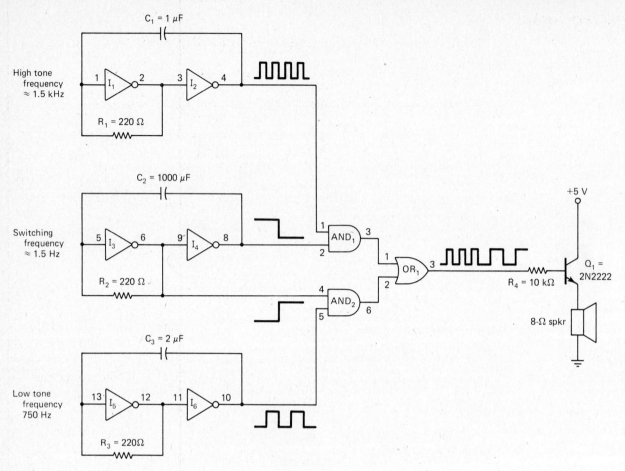

**Figure 4-13**  Two-tone alarm circuit.

- *1*   1-$\mu$F capacitor at 10 WV dc ($C_1$)
- *1*   2-$\mu$F capacitor at 10 WV dc ($C_3$)
- *1*   1000-$\mu$F capacitor at 10 WV dc ($C_2$)
- *1*   2N2222 transistor or equivalent ($Q_1$)
- *1*   8-$\Omega$ speaker
- *1*   Dual-trace oscilloscope
      Several hookup wires or leads

*Procedure:*

1. Construct the circuit shown in Figure 4-13. A two-tone sound should be heard from the speaker.

2. With the oscilloscope, measure the output of $I_2$ (pin 4), calculate the frequency, and record here: high tone = _____ Hz.

3. With the oscilloscope, measure the output of $I_6$ (pin 10), calculate the frequency, and record here: low tone = _____ Hz.

4. With the oscilloscope look at the switching action at the output of $I_3$ (pin 6) and $I_4$ (pin 8).

5. Place the oscilloscope at the output of OR$_1$ (pin 3) and observe the switching tones that are being fed to $Q_1$.

*Fill-In Questions:*

1. Referring to Figure 4-13, when pin 2 of AND$_1$ is high, the _____

   _____ tone can be heard at the speaker.

2. Referring to Figure 4-13, when pin 4 of AND$_2$ is high, the _____

   _____ tone can be heard at the speaker.

3. Both high and low tones are sent through gate _____ .

4. Using an oscilloscope at the output of OR$_1$, the two tones can be seen

   _____ back and forth.

5. The clock using inverters _____ and _____ determines

   the rate of switching for the two tones.

## EXPERIMENT 6. 555 TIMER ONE-SHOT TRIGGER CIRCUIT

*Objective:*

To demonstrate the operation of a 555 timer one-shot (monostable multi-vibrator) trigger circuit and show how to calculate the duration of the output voltage pulse.

*Introduction:*

Figure 4-14a shows the basic 555 timer one-shot trigger circuit. Resistors $R_1$ and $R_2$ serve only as a method for triggering the circuit into operation. Components $R_A$ and $C_1$ determine the width of the output-voltage pulse.

**Figure 4-14** 555 timer IC one-shot trigger circuit: (a) circuit; (b) data table. (From F. Hughes, *Basic Electronics: Theory and Experimentation,* Prentice-Hall, Englewood Cliffs, N.J., © 1984, Fig. 12-9, p. 277. Reprinted with permission.)

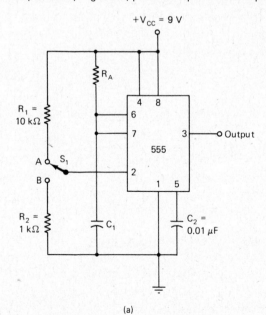

| $R_A$ ($\Omega$) | $C_1$ ($\mu$F) | $T = 1.1\,R_A C_1$ calculated (ms) | $T$ measured |
|---|---|---|---|
| 1 M | 10 | | |
| 470 k | 10 | | |
| 100 k | 10 | | |
| 10 k | 10 | | |
| 470 k | 1 | | |
| 470 k | 0.1 | | |

(a)                                                              (b)

*Materials Needed:*

*1*   +5-V to +15-V power supply

*1*   Standard or digital voltmeter

*1*   Dual-trace oscilloscope

*1*   555 timer IC

*1*   1-k$\Omega$ resistor at 0.5 W ($R_2$)

*2*   10-k$\Omega$ resistors at 0.5 W ($R_1$ and $R_A$)

*1*   100-k$\Omega$ resistor at 0.5 W ($R_A$)

*1*   470-k$\Omega$ resistor at 0.5 W ($R_A$)

*1*   1-M$\Omega$ resistor at 0.5 W ($R_A$)

*1*   0.01-$\mu$F capacitor at 25 WV dc ($C_2$)

*1*   0.1-$\mu$F capacitor at 25 WV dc ($C_1$)

*1*   1-$\mu$F capacitor at 25 WV dc ($C_2$)

*1*   10-$\mu$F capacitor at 25 WV dc ($C_1$)

*1*   SPDT switch

*1*   Breadboard for constructing circuit

     Several hookup wires or leads

*Procedure:*

1. Construct the circuit shown in Figure 4-14a, using the values given in the first row of the data table of Figure 4-14b for $R_A$ and $C_1$.
2. Connect channel 1 of the oscilloscope at pin 2 and channel 2 at pin 3 to make voltage measurements.
3. Calculate the output voltage pulse width and record in the data table using the formula $T = 1.1 R_A C_1$.
4. Momentarily move $S_1$ to position $B$ and then back to position $A$ and attempt to measure the output pulse width with a watch. Approximate the measured output pulse width and record it in the data table.
5. Initiate another timing cycle using $S_1$ and while the output is high, operate $S_1$ a few times to see that it will not interfere with the timing cycle set by $R_A$ and $C_1$. (This timing cycle will last about 11 s.)
6. Repeat steps 3 and 4 for the values of $R_A$ and $C_1$ given in the data table.

*Fill-In Questions:*

1. For a 555 timer one-shot trigger circuit, the time the output is high is

   determined by _____ and _____.

2. Once the one-shot circuit is triggered, other trigger pulses will not

   _____ with the timing cycle.

3. The one-shot trigger circuit can be used to _____, _____

   _____, and as _____ (use statements).

## EXPERIMENT 7.  74121 IC ONE-SHOT TRIGGER CIRCUIT

*Objective:*

To show the operation of the special 74121 monostable multivibrator DIP IC.

*Introduction:*

This experiment shows how to change the timing cycle of the 74121, the use of positive and negative triggering, and how a timing cycle cannot be interrupted by subsequent pulses. Referring to Figure 4-15a, a positive-going or logic 1 pulse can replace $R_2$ and $S_1$. Similarly, in Figure 4-15b, a negative-going or logic 0 pulse can replace $R_1$ and $S_1$.

*Materials Needed:*

1   Digital logic trainer *or*
1   74121 monostable multivibrator DIP IC
1   1-k$\Omega$ resistor at 0.5 W ($R_2$)
1   10-k$\Omega$ resistor at 0.5 W ($R_1$)
1   100-k$\Omega$ resistor at 0.5 W ($R_1$)
1   1-$\mu$F capacitor at 10 WV dc ($C_1$)
1   10-$\mu$F capacitor at 10 WV dc ($C_1$)
1   100-$\mu$F capacitor at 10 WV dc ($C_1$)
1   Dual-trace oscilloscope
1   Square-wave generator with a frequency up to 100 kHz
    Several hookup wires or leads

*Procedure:*

1.  Construct the circuit shown in Figure 4-15a, using the values given in the first row of the data table.
2.  Connect the oscilloscope to the $Q$ and $\overline{Q}$ outputs. Adjust the two traces so that they are evenly spaced on the face of the cathode ray tube. Set the vertical amplitude selector switch to 1 V/cm or 5 V/cm if possible.
3.  Calculate the duration of the timing cycle with the formula $T = 0.693R_1C_1$ and record in the third column of the data table.
4.  Momentarily move $S_1$ to position $B$ and then back to position $A$. Notice that the oscilloscope trace connected to the $Q$ output rises and the trace connected to the $\overline{Q}$ output falls.
5.  Using the second hand on a watch, try to time the duration of the change at outputs $Q$ and $\overline{Q}$.
6.  Repeat steps 4 and 5 for the remaining values of $R_1$ and $C_1$ in the data table.
7.  Modify the circuit as shown in Figure 4-15b.
8.  Operate $S_1$ to see how negative triggering can be used with the 74121 IC.
9.  Construct the circuit shown in Figure 4-15c.

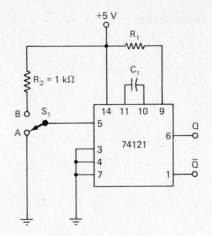

| $R_1$ (kΩ) | $C_1$ (μF) | $T = 0.693R_1C_1$ (sec) |
|---|---|---|
| 100 | 100 | |
| 100 | 10 | |
| 100 | 1 | |
| 10 | 100 | |
| 10 | 10 | |

(a)

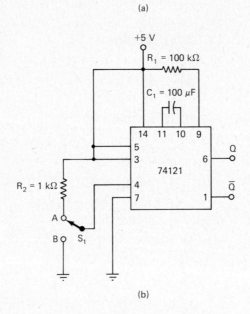

(b)

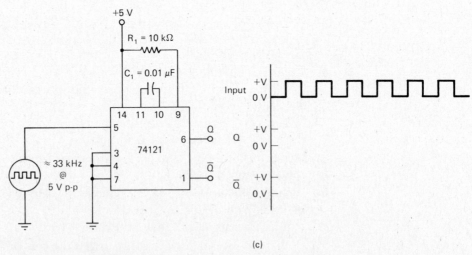

(c)

**Figure 4-15**  74121 one-shot trigger circuit: (a) positive triggering; (b) negative triggering; (c) timing cycle analysis.

10. Connect the oscilloscope to the $Q$ and $\overline{Q}$ outputs.

11. Set the square-wave generator at about 33 kHz with an amplitude of 5 volts peak-to-peak.

12. Observe the output waveforms on the oscilloscope. Notice that during the timing cycle of the 74121, two or more trigger pulses appear at the input, but the circuit will not respond to them.

13. Draw the voltage waveforms of the $Q$ and $\overline{Q}$ outputs on the graph shown in Figure 4-15c. Align these waveforms with the input pulses given.

14. Using the oscilloscope, list the time period of one input pulse here:

    1 input pulse = _____ $\mu$s.

15. Calculate the timing cycle period and list it here: timing cycle = _____

    _____ $\mu$s.

*Fill-In Questions:*

1. The 74121 can be triggered by a _____ - or _____ - going input pulse.

2. The timing cycle of the 74121 depends on the values of _____ and _____.

3. The duration of the timing cycle can be found by the formula _____ _____.

4. Once the monostable multivibrator is triggered, other input pulses will not interfere with its operation until its _____ _____ is complete.

**EXPERIMENT 8. THE SCHMITT TRIGGER CIRCUIT**

*Objective:*

To demonstrate how voltage levels operate the Schmitt trigger inverter circuit.

*Introduction:*

The first part of this experiment establishes the input threshold voltages with a dc circuit. The second part of the experiment shows the effects of ac on the Schmitt trigger inverter circuit. The diode in Figure 4-16c rectifies the ac signal, since negative voltages applied to the inputs of TTL ICs may destroy them.

*Materials Needed:*

1   Digital logic trainer *or*
1   7413 dual Schmitt trigger DIP IC *or*
1   7414 hex Schmitt trigger DIP IC *or*

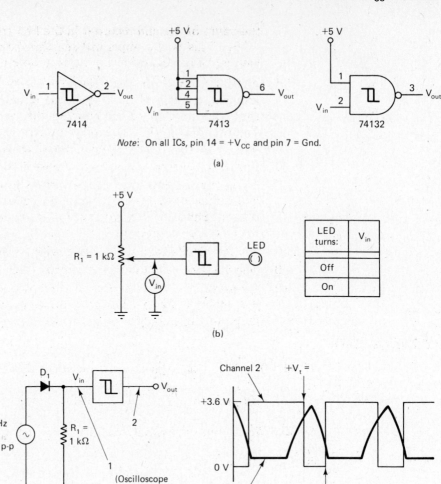

*Note*: On all ICs, pin 14 = +$V_{CC}$ and pin 7 = Gnd.

(a)

(b)

(c)

**Figure 4-16** Schmitt triggers: (a) pin numbers for various ICs; (b) dc level test circuit; (c) ac test circuit.

*1*   74132 quad Schmitt trigger DIP IC
*1*   1N4001 diode or equivalent ($D_1$)
*1*   1-k$\Omega$ resistor at 0.5 W ($R_1$)
*1*   1-k$\Omega$ potentiometer ($R_1$)
*1*   LED indicator
*1*   Dual-trace oscilloscope
*1*   Sine-wave signal generator
*1*   Voltmeter
      Several hookup wires or leads

*Procedure:*

1. Construct the circuit shown in Figure 4-16b. Modify the Schmitt trigger as shown in Figure 4-16a, depending on which IC is being used.
2. Place the wiper of $R_1$ at ground ($V_{in} = 0$ V). The LED should be on.
3. Using $R_1$, slowly begin to increase $V_{in}$. When the LED turns off, notice

the value of $V_{in}$ and record it in the first row of the data table in Figure 4-16b. This is the upper voltage threshold ($+V_t$). Steps 2 and 3 may have to be performed a few times to obtain an accurate reading.

4. Set the wiper of $R_1$ for a $V_{in}$ of +3 V. The LED should be off.

5. Using $R_1$, slowly begin to decrease $V_{in}$. When the LED turns on, notice the value of $V_{in}$ and record it in the second row of the data table in Figure 4-16b. This is the lower voltage threshold ($-V_t$). Steps 4 and 5 may have to be performed a few times to obtain an accurate reading.

6. Construct the circuit shown in Figure 4-16c.

7. Using the oscilloscope, set the vertical input selector switch to 1 V/cm and adjust both traces to the same reference level (one on top of the other). This is called *superimposing signals.* Connect the oscilloscope to the circuit as indicated.

8. Set the signal generator at 1 kHz with an amplitude of 5 V peak-to-peak.

9. The waveforms should look similar to the graph shown in Figure 4-16c.

10. From the waveforms on the oscilloscope, determine the voltages of $+V_t$ and $-V_t$ and record these values in their appropriate place on the graph of Figure 4-16c.

*Fill-In Questions:*

1. The Schmitt trigger circuit output will change states depending on specific voltage _____ at the inputs.

2. When $+V_t$ is reached, the output of the Schmitt trigger inverter circuit will be _____.

3. When $-V_t$ is reached, the output of the Schmitt trigger inverter circuit will be _____.

**SECTION 4-5
INSTANT REVIEW**

• The clock in digital circuits is an astable multivibrator that has no stable state, but switches back and forth between two states, providing a square-edge voltage output. This oscillator, as it is sometimes called, may have a single output or a complementary $Q$ and $\overline{Q}$ output. The frequency of the clock is usually determined by $RC$ time-constant components or sometimes by a crystal.

• A monostable multivibrator is a trigger circuit that has only one stable state. It can be triggered to change to the other state for a predetermined time, depending on the $RC$ time constant, after which it returns to the original state. A one-shot multivibrator can be used for reshaping pulses, stretching or delaying pulses, debouncing mechanical switches, and providing a sharp-leading-edge pulse for other circuits.

• The Schmitt trigger circuit changes its output state depending on the threshold level of voltages at its input. The upper and lower threshold voltages set the hysteresis loop.

**SECTION 4-6**
**EXERCISES II**

Perform all the exercises in this section before beginning the next section.

1. Find the output pulse duration for a 555 timer one-shot trigger circuit for the following values. Use the formula $T = 1.1R_AC_1$.

   a. $R_A = 1\ k\Omega$
      $C_1 = 0.01\ \mu F$

      $T = $ _____ s

   b. $R_A = 1\ k\Omega$
      $C_1 = 0.002\ \mu F$

      $T = $ _____ s

   c. $R_A = 22\ k\Omega$
      $C_1 = 0.1\ \mu F$

      $T = $ _____ s

   d. $R_A = 100\ k\Omega$
      $C_1 = 2\ \mu F$

      $T = $ _____ s

2. Find the output pulse duration for a 74121 one-shot trigger circuit for the following values. Use the formula $T_p = 0.693R_1C_1$.

   a. $R_1 = 47\ k\Omega$
      $C_1 = 10\ \mu F$

      $T_p = $ _____ s

   b. $R_1 = 22\ k\Omega$
      $C_1 = 1\ \mu F$

      $T_p = $ _____ s

   c. $R_1 = 33\ k\Omega$
      $C_1 = 0.02\ \mu F$

      $T_p = $ _____ s

   d. $R_1 = 10\ k\Omega$
      $C_1 = 0.5\ \mu F$

      $T_p = $ _____ s

3. List the input threshold voltages for a standard TTL Schmitt trigger circuit.

   a. $+V_t = $ _____

   b. $-V_t = $ _____

4. Indicate the output condition (low or high) for a TTL Schmitt trigger inverter for the input threshold voltages given.

   a. input $= +V_t$, output $= $ _____

   b. input $= -V_t$, output $= $ _____

5. List the types of clocks under the heading that describes them best. Use: inverter clock, NAND gate clock, NOR gate clock, crystal-controlled clock, and 555 timer clock.

   *Distorted Output Waveform*          *Good Square-Wave Output*          *Good Frequency Stability*

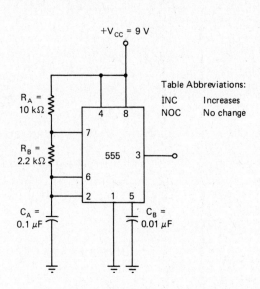

| Condition | $V_{out}$ (dc) | $v_{out}$ (signal) | Comments |
|---|---|---|---|
| Normal | | | All voltages are proper |
| $R_A$ open | | INC | No charge path for $C_A$; output saturates |
| $R_B$ open | | INC | No charge path for $C_A$; output saturates |
| $C_A$ open | | INC | Wiring capacitance causes high frequency ($\approx$ 200 kHz) |
| $C_A$ shorted | | INC | No charging; output saturates |
| $C_B$ open | | NOC | No apparent trouble; check for noise or 60-Hz modulation |
| $C_B$ shorted | | NOC | Pulse width narrows and signal seems to invert |

**Figure 4-17** 555 clock: (a) circuit; (b) data table. (From F. Hughes, *Basic Electronics: Theory and Experimentation,* Prentice-Hall, Englewood Cliffs, N.J., © 1984, figure on p. 286. Reprinted with permission.)

## SECTION 4-7
## TROUBLESHOOTING APPLICATION:   555 TIMER IC CLOCK

Construct the circuit shown in Figure 4-17. Open or short the components as shown in the data table and record the voltages in the proper place. The abbreviations will help indicate the voltage conditions associated with each problem. All voltages are referenced to ground.

## SECTION 4-8
## SELF-CHECKING QUIZZES

### 4-8a   CLOCKS AND TRIGGER CIRCUITS: TRUE-FALSE QUIZ

Place a T for true or an F for false to the left of each statement.

_____ 1.  A clock may have one or two outputs.

_____ 2.  A crystal-controlled clock is not very stable.

_____ 3.  When the $Q$ output is low, the $\overline{Q}$ output will be high.

_____ 4.  Trigger circuit timing components usually consist of resistors and capacitors.

_____ 5.  NAND/NOR gate clocks are not as reliable as an inverter clock.

_____ 6.  The 555 timer IC can be used as a clock or a one-shot trigger circuit.

_____ 7.  The 74121 can be triggered by a positive-going or a negative-going pulse.

_____ 8.  A one-shot trigger circuit can be triggered at any time.

_____ 9.  The threshold voltages for a TTL Schmitt trigger circuit are $+V_t$ = 0.9 V and $-V_t$ = 1.7 V.

_____ 10.  The duty cycle is a comparison of the on and off times of a device.

## 4-8b  CLOCKS AND TRIGGER CIRCUITS: MULTIPLE-CHOICE QUIZ

Circle the correct answer for each question.

1.  The operation of a clock is known as:

    a.  monostable

    b.  bistable

    c.  astable

    d.  quasi-stable

2.  The operation of a one-shot trigger circuit is known as:

    a.  monostable

    b.  bistable

    c.  astable

    d.  quasi-stable

3.  A basic crystal-controlled clock has:

    a.  poor stability

    b.  good stability

    c.  a sawtooth wave output

    d.  none of the above

4.  The voltage levels from +0.8 V to +2.0 V for TTL digital circuits is defined as:

    a.  logic 1

    b.  forbidden area

    c.  logic 0

    d.  none of the above

5.  The IC that has internal circuits that enable the circuit's output to change states at specific voltage levels without external biasing is the:

    a.  7400 NAND gate IC

    b.  7404 hex inverter IC

    c.  7413 Schmitt trigger IC

    d.  555 timer IC

6.  The best clock to use in a digital system requiring a frequency of 800 kHz is the:

    a.  inverter clock

    b.  NAND/NOR clock

    c.  555 timer clock

    d.  crystal-controlled clock

Questions 7 and 8 refer to Figure 4-18a.

7.  The frequency of the 555 timer is about:

    a.  10 kHz

    b.  15 kHz

    c.  17 kHz

    d.  23 kHz

8.  The timing cycle of the 74121 is about:

    a.  30 $\mu$s

    b.  229 $\mu$s

    c.  60 $\mu$s

    d.  330 $\mu$s

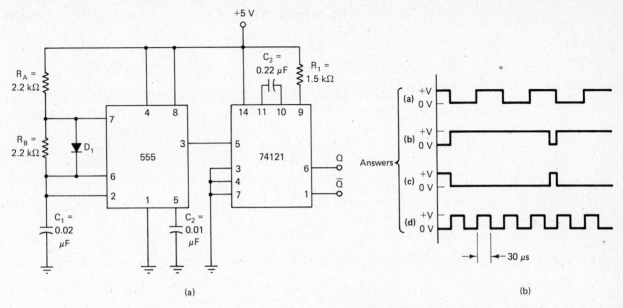

(a)

(b)

**Figure 4-18**

Use the waveform graph in Figure 4-18b for the answers to questions 9 and 10.

9. The voltage waveform of the output for the 555 timer IC is:

   a.

   b.

   c.

   d.

10. The voltage waveform of the $\overline{Q}$ output of the 74121 IC is:

   a.

   b.

   c.

   d.

## ANSWERS TO EXPERIMENTS AND QUIZZES FOR UNIT 4

*Experiment 1.*

      (1) square    (2) decrease    (3) decreased

*Experiment 2.*

      (1) inverter    (2) square    (3) $f = 1/3R_1C_1$

*Experiment 3.*

      (1) crystal    (2) stable    (3) not

*Experiment 4.*

      (1) $R_A$, $R_B$, $C_1$    (2) decreases    (3) increases    (4) unsymmetrical, less
      (5) symmetrical, 50    (6) 5, 95

*Experiment 5.*

      (1) high    (2) low    (3) $OR_1$    (4) switching    (5) $I_3$, $I_4$

*Experiment 6.*

      (1) $R_A$, $C_1$    (2) interfere    (3) reshape a pulse, stretch a pulse, a bounce-less switch

*Experiment 7.*

(1) negative, positive      (2) $R_1, C$      (3) $T = 0.693 R_1 C_1$      (4) timing cycle

*Experiment 8.*

(1) levels      (2) low      (3) high

*True-False:*

(1) T      (2) F      (3) T      (4) T      (5) F      (6) T      (7) T      (8) F      (9) F
(10) T

*Multiple-Choice:*

(1) c      (2) a      (3) b      (4) b      (5) c      (6) d      (7) c      (8) b      (9) d
(10) c

# Unit 5

---

# Flip-Flops

**SECTION 5-1**
**IDENTIFICATION, THEORY, AND OPERATION**

**5-1a  INTRODUCTION**

Flip-flops are bistable multivibrators that can remain in one of two conditions or states. A flip-flop may have one, two, or three inputs and usually two ouputs ($Q$ and $\overline{Q}$). It may take one or two inputs to turn a flip-flop on and off. A flip-flop is considered off when its $Q$ output is 0 and its $\overline{Q}$ output is 1. When it is on, outputs $Q = 1$ and $\overline{Q} = 0$.

Logic gates are used in combinational logic, which was studied in Units 2 and 3. Flip-flops are used in *sequential logic*, which involves timing and memory devices. The flip-flop is a basic memory device, since it can be turned on and remembers that it is on after the control pulses are removed. This holding or locking action in a particular state is called *latching* and often a flip-flop is referred to as a *latch*. Flip-flops are wired together to form shift registers, counters, and various memory circuits which will be studied in the following units.

**5-1b  *R-S* FLIP-FLOP**

An $R$-$S$ flip-flop has two inputs, the set input ($S$) for turning it on and the reset input ($R$) for turning it off. It has a $Q$ and $\overline{Q}$ output. Two NAND gates can be wired together to produce an $R$-$S$ flip-flop as shown in Figure 5-1a. Remember, a NAND gate will produce a 1 at its output when any input is 0. The inputs are normally high or a 1. If the flip-flop is considered reset (off), then output $Q = 0$ and output $\overline{Q} = 1$. In this case the output of $ND_1$ is 0 and applied back to the input of $ND_2$, which keeps its output at a 1. This 1 is applied back to the input of $ND_1$, which keeps its output at 0, hence the

133

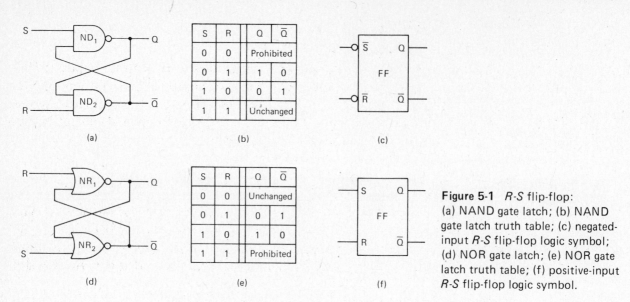

**Figure 5-1** *R-S* flip-flop:
(a) NAND gate latch; (b) NAND
gate latch truth table; (c) negated-
input *R-S* flip-flop logic symbol;
(d) NOR gate latch; (e) NOR gate
latch truth table; (f) positive-input
*R-S* flip-flop logic symbol.

latch condition. With both inputs a 1, the output will remain unchanged whether the flip-flop is on or off. This is the remembering or storage state of the device.

If both inputs were 0, both outputs would be 1, which is prohibited for a flip-flop since the output must be complementary. The following sequence of operations shows what happens to turn on and turn off this NAND gate latch and can be verified by the truth table in Figure 5-1b.

*To Turn On:*

1. The $S$ input goes to 0.
2. The output of $ND_1$ goes to 1, which is applied back to the input of $ND_2$.
3. The output of $ND_2$ goes to 0, which is applied back to the input of $ND_1$.
4. The $S$ input goes to 1, but the output of $ND_1$ does not change state because there is a 0 at the other input. The latch is on and considered storing a 1 at the $Q$ output.

*To Turn Off:*

1. The $R$ input goes to 0.
2. The output of $ND_2$ goes to 1, which is applied back to the input of $ND_1$.
3. The output of $ND_1$ goes to 0, which is applied back to the input of $ND_2$.
4. The $R$ input goes to 1, but the output of $ND_2$ does not change state because there is a 0 at the other input. The latch is off and considered storing a 0 at the $Q$ output.

Figure 5-1c shows the logic symbol for a negated-input $R-S$ flip-flop. The bubbles at the inputs indicate that it takes 0s or negative-going pulses to initiate the proper action of the device.

Two NOR gates can also be wired to produce an $R-S$ flip-flop, as shown in Figure 5-1d. Remember, a NOR gate will produce a 1 at its output only when all inputs are 0. Normally, in the storing state, the inputs are at 0 and the outputs will be unchanged. The following sequence of operations shows what happens to turn this NOR gate latch on and off and can be verified by the truth table in Figure 5-1e. Assume that the latch is off and $Q = 0$, $\overline{Q} = 1$.

*To Turn On:*

1. The $S$ input goes to 1.
2. The output of $NR_2$ goes to 0 and is applied back to the input of $NR_1$.
3. The output of $NR_1$ goes to 1, which is applied back to the input of $NR_2$.
4. The $S$ input goes to 0, but the output of $NR_2$ does not change state because there is a 1 at the other input. The latch is on and considered storing a 1 at its $Q$ output.

*To Turn Off:*

1. The $R$ output goes to 1.
2. The output of $NR_1$ goes to 0 and is applied back to the input of $NR_2$.
3. The output of $NR_2$ goes to 1, which is applied back to the input of $NR_1$.
4. The $R$ input goes to 0, but the output of $NR_1$ does not change states because there is a 1 at the other input. The latch is off and considered storing a 0 at its $Q$ output.

It is prohibited to have both $S$ and $R$ inputs 1 at the same time for the NOR gate latch. Figure 5-1f shows the logic symbol for this latch. These two types of flip-flops need only one input to turn on and one input to turn off. This is referred to as *asynchronous operation*, since no other timing pulses are needed.

## 5-1c  THE BOUNCELESS SWITCH

When a mechanical switch is closed, its contacts bounce or vibrate a minute amount before the switch comes to rest in the new position. Actually, the contacts open and close a few times, which can cause extraneous pulses, as shown in Figure 5-2a. In digital circuits, these extra pulses can false-trigger other circuits and cause erratic operation. If the switch is used with a NAND gate latch, called a *bounceless switch circuit*, shown in Figure 5-2b, the out-

**Figure 5-2** Bounceless switch: (a) extraneous pulses caused by switch contact bounding; (b) NAND gate latch to eliminate extraneous pulses caused by switch contact bounce.

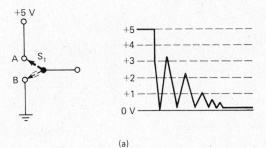

(a)

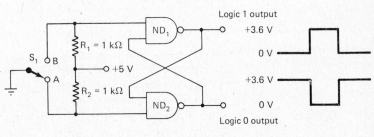

(b)

puts will not respond to the fast bouncing and will remain in the state of the first or initial pulse from the switch. With the position of $S_1$ at ground as shown, a 0 is at the input of $ND_2$, which keeps its output at 1. This 1 is applied back to one input of $ND_1$. The other input of $ND_1$ is pulled up through $R_1$ toward +5 V, a 1 also, and the output is 0. When $S_1$ is moved from position $A$ to position $B$, this input to $ND_1$ is now grounded or at 0 and the output goes to 1. This 1 is applied back to one input of $ND_2$ and since its other input is no longer grounded, it is pulled up through $R_2$ to +5 V or a 1, which causes the output to go to 0.

The outputs will produce a sharp-leading-edge pulse, one going positive and the other going negative. They are labeled logic 1 and logic 0, which indicates the states they produce when $S_1$ is activated.

### 5-1d CLOCKED *R-S* FLIP-FLOP

The NAND gate latch is provided with two other NAND gates, as shown in Figure 5-3a, to produce the clocked *R-S* flip-flop. This type of flip-flop needs a clock pulse every time to make it change states, depending on the condition of the $S$ and $R$ inputs. Assuming that the latch is off ($Q = 0$, $\overline{Q} = 1$), the following sequence of operations shows what happens to turn this flip-flop on and off. Inputs $S$, $R$, and Clk are initially 0 (the flip-flop is in the storing state and assumed to be off).

*To Turn On:*

1. The $S$ input goes to 1, but nothing happens.
2. The Clk input goes to 1 and the output of $ND_1$ goes to 0, which causes the output of $ND_3$ to go to a 1, which is applied back to one input of $ND_4$. Since input $R$ is at 0 the output of $ND_2$ is 1, so now the output of $ND_4$ goes to 0, and is applied back to one input of $ND_3$.
3. The $S$ input and Clk input go to 0, and the output is storing a 1 at $Q$ and a 0 at $\overline{Q}$.

*To Turn Off:*

1. The $R$ input goes to 1, but nothing happens.
2. The Clk input goes to 1 and the output of $ND_2$ goes to 0. This causes the output of $ND_4$ to go to a 1, which is applied back to one input of $ND_3$. Since input $S$ is 0, the output of $ND_1$ is 1, so now the output of $ND_3$ goes to 0 and is applied back to one input of $ND_4$.
3. The $S$ input and the Clk input go to 0, and the output of the flip-flop is storing a 0 at $Q$ and a 1 at $\overline{Q}$.

**Figure 5-3** Clocked *R-S* flip-flop: (a) NAND gate circuit; (b) truth table; (c) logic symbol.

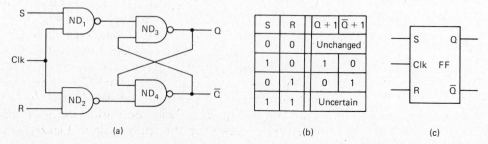

| S | R | Q + 1 | $\overline{Q}$ + 1 |
|---|---|-------|--------------------|
| 0 | 0 | Unchanged | |
| 1 | 0 | 1 | 0 |
| 0 | 1 | 0 | 1 |
| 1 | 1 | Uncertain | |

(a)                              (b)                              (c)

The logic table shown in Figure 5-3b verifies the action just described. The labeling of the outputs in the table, $Q + 1$ and $\overline{Q} + 1$, indicates the states of $Q$ and $\overline{Q}$ after a clock pulse appears, depending on the conditions at the $S$ and $R$ inputs. If inputs $S$ and $R$ are both 1 when a clock pulse appears, the outputs may switch to either state and are uncertain; therefore, this is a prohibited condition. Because a clock pulse is needed with this flip-flop to change states, its operation is termed *synchronous* (or needs timing pulses). This type of flip-flop operates with 1s at the input, as indicated by the logic symbol shown in Figure 5-3c.

### 5-1e  TOGGLE (T-TYPE) FLIP-FLOP

The *toggle (T-type) flip-flop*, sometimes called the *complementary flip-flop*, has one input, as shown in Figure 5-4a. Notice that the outputs $Q$ and $\overline{Q}$ are cross-connected back to the input NAND gates $ND_1$ and $ND_2$. This establishes a condition where the outputs steer the inputs for a change in state when a pulse appears at the $T$ input. Because of the propagation delay time (the time it takes the pulses to travel through the circuitry of the gates) the NAND gate T-type flip-flop is unstable and not very practical. Therefore,

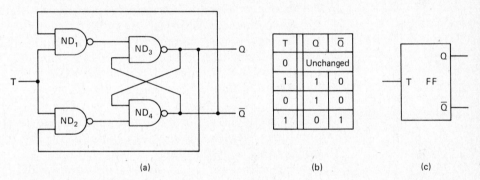

(a)                                          (b)                      (c)

**Figure 5-4**  Toggle (T-type) flip-flop: (a) NAND gate circuit; (b) truth table; (c) logic symbol.

a sequence of operations will not be given here; however, a T-type flip-flop will change states (or the outputs will complement each other) when a pulse appears at the input. The truth table shown in Figure 5-4b illustrates its operation. Notice that the logic symbol in Figure 5-4c shows the flip-flop to be positive-triggered, whereas in actual practice many T-type flip-flops are triggered by negative-going pulses and are indicated with a bubble at the input. These types of flip-flops are used quite extensively in digital circuits.

### 5-1f  *J-K* FLIP-FLOP

Flip-flops are connected together to form shift registers and counters. At high speeds of operation the pulses race through the circuits and inputs may respond falsely to output conditions, creating a problem in accuracy. The *J-K* flip-flop was designed to reduce or eliminate this "race" problem.

Figure 5-5a shows a functional NAND gate *J-K* flip-flop. Notice that the two input NAND gates ($ND_1$ and $ND_2$) have three inputs. The added $J$ input is similar to the $S$ input and the $K$ input is similar to the $R$ input of a clocked $R$-$S$ flip-flop, and the normal turn-on and turn-off operation is the same, as shown by the truth table in Figure 5-5b. Like the T-type flip-

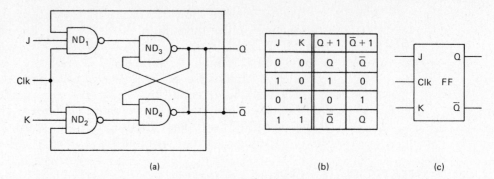

**Figure 5-5**  *J-K* flip-flop: (a) NAND gate circuit; (b) truth table; (c) logic symbol.

flop that uses NAND gates, this circuit is not very practical; however, the outputs control the inputs similarly. This means that the inputs cannot turn on the flip-flop unless the output is completely off, and vice versa. A basic *J-K* flip-flop logic symbol is shown in Figure 5-5c.

A commercial *J-K* flip-flop is a specially designed IC to operate in circuits and may have added inputs as shown in Figure 5-6a. In this form it becomes a universal digital device capable of various operations. The preset (Ps) input can override any normal input pulses at the *J*, *K*, and Clk inputs, and when this lead is brought low (0), the flip-flop will turn on. Similarly, the clear (Clr) or reset input can override any normal inputs, and when this lead is brought low (0) the flip-flop will turn off. These two leads can be used to make this universal flip-flop act like a negated-input *R-S* flip-flop.

This flip-flop uses a negated input for the clock and the symbol $>$ means it is edge-triggered, where the outputs will change state only when the clock pulse is falling. If the *J* and *K* inputs are connected to a logic 1, then each time a negative-going clock pulse arrives the outputs will change state, similar to a T-type flip-flop.

The basic *J-K* flip-flop was improved and functionally has a *master flip-flop* which controls a *slave flip-flop*, as shown in Figure 5-6b. In this case the master flip-flop operates on the leading positive edge of the clock pulse and passes its information at $M_Q$ and $M_{\overline{Q}}$ to the inputs $S_J$ and $S_K$ of the slave flip-flop. The slave flip-flop has a bubble at its clock input, so it operates with the information from the master flip-flop on the trailing negative edge of the same clock pulse. The timing chart shown in Figure 5-6c illustrates the operation of the *master/slave flip-flop*.

A typical dual *J-K* master/slave flip-flop DIP IC is shown in Figure 5-6d. Study the pin identification to see which pins are connected to each flip-flop in the package.

## 5-1g  D-TYPE FLIP-FLOP

If an inverter is added from the *J* input to the *K* input of a *J-K* flip-flop as shown in Figure 5-7a, it becomes a data or delay (D-type) flip-flop, as shown in Figure 5-7b. When the *J* input (*D* input) is 0, the *K* input will be a 1, and when a negative-going clock pulse arrives, the flip-flop will turn off or remain off. If the *J* input (*D* input) is a 1 when the clock pulse arrives, the flip-flop turns on. If the *J* input is 0 when the next clock pulse arrives, the flip-flop will turn off. In effect, the data at the *D* input have been delayed one clock pulse time. In other words, the *Q* output will follow the data present on the *D* input and at least store it for one clock pulse time. The

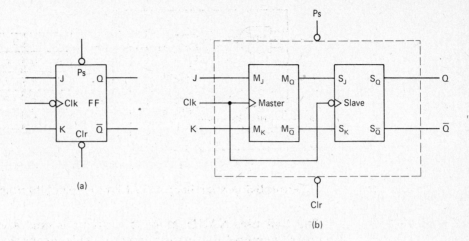

(a)

(b)

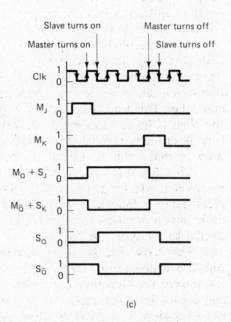

(c)

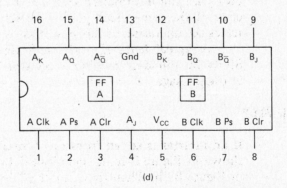

(d)

**Figure 5-6** Commercial *J-K* flip-flop: (a) with added negated clock and preset and clear leads; (b) function diagram for master/slave flip-flop; (c) timing chart for master/slave flip-flop; (d) typical 7476 dual *J-K* master/slave flip-flop IC.

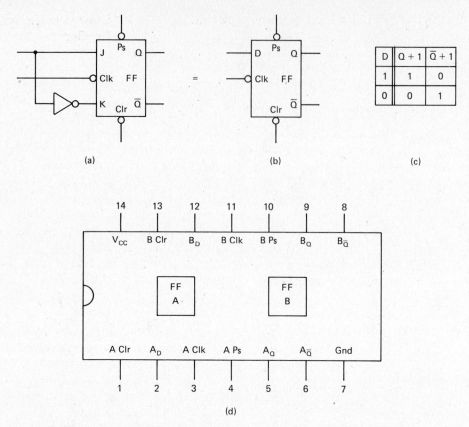

**Figure 5-7** Data or delay (D-type) flip-flop: (a) *J-K* flip-flop with an inverter to *K* input; (b) D-type flip-flop logic symbol; (c) truth table; (d) typical 7474 dual D-type flip-flop DIP IC.

truth table shown in Figure 5-7c illustrates the action of the *D*-type flip-flop. A typical 7474 dual D-type flip-flop DIP IC is shown in Figure 5-7d. Note the pin identification and be able to determine which pins are connected to each flip-flop in the package.

## SECTION 5-2
## EXERCISES I

Perform all the exercises in this section before beginning the next section.

1. Draw the logic symbol for a negated-input *R-S* flip-flop.

2. Draw the symbol for a positive input *R-S* flip-flop.

3.   Draw the circuit for a bounceless switch.

4.   Draw the logic symbol for a clocked *R-S* flip-flop.

5.   Draw the logic symbol for a T-type flip-flop.

6.   Draw the logic gate for a *J-K* flip-flop with negated Clk, Ps, and Clr leads.

7.   Draw the functional diagram for a *J-K* master/slave flip-flop.

8.   Draw the logic symbol for a D-type flip-flop.

9.   In Table 5-1, using 1s and 0s, indicate the input conditions needed to turn on the flip-flops and show their output states. Also show the input conditions needed to turn off the flip-flops and show their output states. Let a logic 0 equal a low or falling pulse and a logic 1 equal a high or rising pulse. Assume that all flip-flops are initially in the off state.

**TABLE 5-1**

| Type of Flip-Flop | To Turn On: | | To Turn Off: | |
|---|---|---|---|---|
| | Inputs | Outputs | Inputs | Outputs |
| R-S | $S =$ ___ , $R =$ ___ | $Q =$ ___ , $\bar{Q} =$ ___ | $S =$ ___ , $R =$ ___ | $Q =$ ___ , $\bar{Q} =$ ___ |
| Clocked R-S | $S =$ ___ , $R =$ ___ , | $Q =$ ___ , $\bar{Q} =$ ___ | $S =$ ___ , $R =$ ___ , | $Q =$ ___ , $\bar{Q} =$ ___ |
| | Clk = ___ | | Clk = ___ | |
| T-type with neg. input | $T =$ ___ | $Q =$ ___ , $\bar{Q} =$ ___ | $T =$ ___ | $Q =$ ___ , $\bar{Q} =$ ___ |
| J-K with neg. Clk | $J =$ ___ , $K =$ ___ , | $Q =$ ___ , $\bar{Q} =$ ___ | $J =$ ___ , $K =$ ___ , | $Q =$ ___ , $\bar{Q} =$ ___ |
| | Clk = ___ | | Clk = ___ | |
| D-type with neg. Clk | $D =$ ___ , | $Q =$ ___ , $\bar{Q} =$ ___ | $D =$ ___ , | $Q =$ ___ , $\bar{Q} =$ ___ |
| | Clk = ___ | | Clk = ___ | |

10. Under each column shown, list the flip-flops according to their type of operation.

　　　　　　　*Asynchronous*　　　　*Synchronous*

**SECTION 5-3**
**DEFINITION EXERCISES**

Give a brief description of each of the following terms.

1. Flip-flop _____

_____

_____

2. *R-S* flip-flop _____

_____

_____

3. Clocked *R-S* flip-flop_____

_____

_____

4. T-type flip-flop_____

_____

_____

5. *J-K* flip-flop_____

_____

_____

6. Master/slave *J-K* flip-flop_____

_____

_____

7. D-type flip-flop_____

_____

_____

8. Asynchronous_____

_____

_____

9. Synchronous_____

_____

_____

10. Bounceless switch_____

_____

_____

11. Latch_____

_____

_____

**12.** Edge-triggered (negative and positive)_____

_____

_____

**13.** Bistable _____

_____

_____

**14.** Sequential logic _____

_____

_____

**15.** Set _____

_____

_____

**16.** Reset _____

_____

_____

**SECTION 5-4**
**EXPERIMENTS**

**EXPERIMENT 1.   NAND GATE *R-S* FLIP FLOP**

*Objective:*

> To show the use of NAND gates in constructing a basic latch, a clocked *R-S* flip-flop, and a bounceless switch.

*Introduction:*

> A basic NAND gate flip-flop uses two gates, as shown in Figure 5-8a. Input *S* turns it on and input *R* turns it off. With the addition of two more NAND gates, as shown in Figure 5-8b, the clocked *R-S* flip-flop is produced. The clock input must have a pulse simultaneously with the *S* or *R* input for proper operation. The NAND gate bounceless switch circuit shown in Figure 5-8c produces sharp output pulses.

*Materials Needed:*

> *1*   Digital logic trainer *or*
> *1*   7400 quad two-input NAND gate DIP IC
> *2*   LED indicators

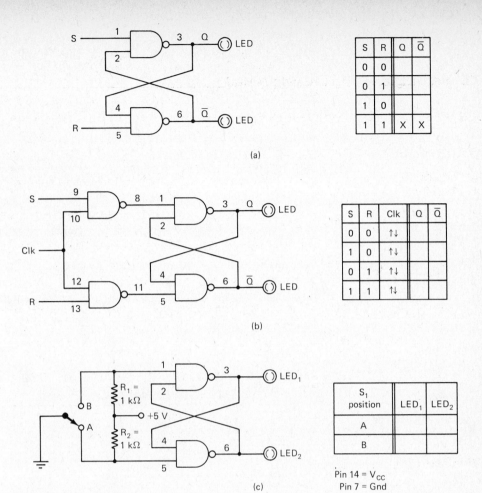

**Figure 5-8** *R-S* flip-flop: (a) NAND gate *R-S* flip-flop and truth table; (b) clocked NAND gate *R-S* flip-flop and truth table; (c) NAND gate bounceless switch circuit.

*2* 1-kΩ resistors at 0.5 W ($R_1$ and $R_2$)

*1* SPDT switch ($S_1$)

Several hookup wires or leads

*Procedure:*

1. Construct the circuit shown in Figure 5-8a. Use logic switches for inputs $S$ and $R$.

2. Set the input switches according to the input conditions in the truth table and record the output $Q$ and $\overline{Q}$ states in the proper place. Remember, an LED that is on = 1 and an LED that is off = 0.

3. Modify the circuit as shown in Figure 5-8b to produce a clocked $R$-$S$ flip-flop.

4. Set the input switches $S$ and $R$ according to this truth table. Then move the clock switch up ↑ and then down ↓.

5. Observe the outputs $Q$ and $\overline{Q}$, and then record their states in the truth table.

6. Construct the circuit shown in Figure 5-8c. Clip leads may be needed to connect the resistors and switch $S_1$ to the logic trainer. *(Do not use a logic switch. $S_1$ is connected directly to ground.)*

7. Set $S_1$ according to the truth table and record the condition of LED$_1$ and LED$_2$ in the proper place.

*Fill-In Questions:*

1.  With an *R-S* flip-flop, having both inputs at 1 is _____.

2.  With a clocked *R-S* flip-flop, a clock pulse is needed _____ time together with either the *S* or *R* input to cause the proper action.

3.  The clocked *R-S* flip-flop is a _____ operation.

4.  A NAND gate bounceless switch has two outputs and when the switch is actuated, one output goes from low to _____ and the other output goes from high to _____.

### EXPERIMENT 2.   THE UNIVERSAL *J-K* FLIP-FLOP

*Objective:*

To demonstrate how the *J-K* flip-flop can be connected to produce an *R-S*, a clocked *R-S*, and a T-type flip-flop.

*Introduction:*

Figure 5-9a shows the connections for testing and experimenting with the *J-K* flip-flop. Figure 5-9b shows an alternative method for constructing a T-type flip-flop using the *J-K* flip-flop.

*Materials Needed:*

1   Digital logic trainer *or*
1   7476 dual *J-K* master/slave flip-flop DIP IC
2   LED indicators
    Several hookup wires or leads

**Figure 5-9**   *J-K* flip-flop: (a) universal circuit; (b) *J-K* flip-flop wired as a T-type flip-flop.

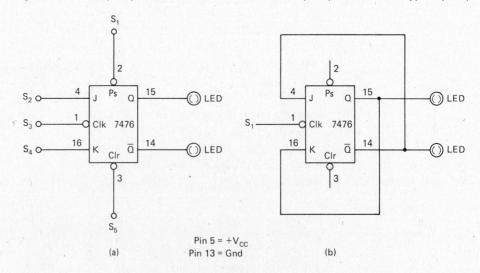

*Procedure:*

1. Connect the circuit as shown in Figure 5-9a. Initially set the logic switches thus: $S_1$, $S_5 = 1$; $S_2$, $S_3$, $S_4 = 0$.
2. Move $S_5$ (Clr) down and up. The flip-flop should now be off ($Q = 0$, $\overline{Q} = 1$).

*R-S Flip-Flop Operation*

3. Move $S_1$ (Ps) down and up. The flip-flop should be on. Record the state of the outputs here: $Q =$ _____, $\overline{Q} =$ _____.

4. Move $S_5$ (Clr) down and up. The flip-flop should be off. Record the state of the outputs here: $Q =$ _____, $\overline{Q} =$ _____.

*Clocked R-S Flip-Flop Operation*

5. Move $S_2$ ($J$) up. Move $S_3$ (Clk) up and down. (Notice that the flip-flop does not turn on until $S_3$ is brought down.) Move $S_2$ ($J$) down. Record the state of the outputs here: $Q =$ _____, $\overline{Q} =$ _____.

6. Move $S_4$ ($K$) up. Move $S_3$ (Clk) up and down. (Again, notice that the flip-flop does not turn off until $S_3$ is brought down.) Move $S_4$ ($K$) down. Record the state of the outputs here: $Q =$ _____, $\overline{Q} =$ _____.

*T-Type Flip-Flop Operation*

7. Move $S_2$ ($J$) and $S_4$ ($K$) up. Move $S_3$ (Clk) up and down. The flip-flop changed states when: $\underline{S_3 \text{ went up}}$, $\underline{S_3 \text{ went down}}$ (circle the correct answer). Record the state of the outputs here: $Q =$ _____, $\overline{Q} =$ _____.

8. Move $S_3$ (Clk) up and down. The flip-flop changed states when: $\underline{S_3 \text{ went up}}$, $\underline{S_3 \text{ went down}}$ (circle the correct answer). Record the state of the outputs here: $Q =$ _____, $\overline{Q} =$ _____. Move $S_2$ ($J$) and $S_4$ ($K$) down.

*Priority of Inputs Ps and Clr*

9. Move $S_5$ (Clr) down. Move $S_2$ ($J$) up. Move $S_3$ (Clk) up and down. Did the flip-flop turn on? _____. Move $S_2$ ($J$) down and $S_5$ (Clr) up.

10. Move $S_1$ (Ps) down. Move $S_4$ ($K$) up. Move $S_3$ (Clk) up and down. Did the flip-flop turn off? _____.

*T-Type Flip-Flop with Outputs Connected to Inputs*

11. Connect the circuit as shown in Figure 5-9b. The outputs are cross-connected to the inputs, $Q$ to $K$ and $\overline{Q}$ to $J$, to form a T-type flip-flop. The Ps (pin 2) and Clr (pin 3) inputs may not need to be wired since unconnected TTL inputs float high or go to a logic 1.

12. Move $S_1$ (Clk) up and down. Did the flip-flop change states? <u>Yes</u>, <u>No</u> (circle one).

13. Repeat step 12 several times to better understand the action of a T-type flip-flop.

*Fill-In Questions:*

1. An *R-S* flip-flop can be made from a *J-K* flip-flop by using only inputs

   _____ and _____.

2. When inputs $J = 1$, $K = 0$, Ps $= 1$, Clr $= 1$, and the Clk is falling, a *J-K*

   flip-flop will turn _____.

3. When inputs $J = 0$, $K = 1$, Ps $= 1$, Clr $= 1$, and Clk is falling, a *J-K* flip-

   flop will turn _____.

4. When inputs $J = 1$, $K = 1$, Ps $= 1$, Clr $= 1$, and Clk is falling, the *J-K*

   flip-flop acts like a _____ flip-flop.

5. When inputs $J = 1$, $K = 0$, Ps $= 1$, Clr $= 0$, and the Clk is falling, the

   outputs $Q =$ _____ and $\overline{Q} =$ _____.

6. When $J = 0$, $K = 1$, Ps $= 0$, Clr $= 1$, and Clk is falling, the outputs $Q =$

   _____ and $\overline{Q} =$ _____.

## EXPERIMENT 3.   D-TYPE FLIP-FLOP

*Objective:*

To show the operation of the D-type flip-flop.

*Introduction:*

The D-type flip-flop shown in Figure 5-10a can be used with normal operations or be used as an *R-S* flip-flop using only inputs Ps and Clr. When it is wired as shown in Figure 5-10b, it becomes a T-type flip-flop. Figure 5-10c shows the use of four D-type flip-flops in a single IC package.

*Materials Needed:*

*1*  Digital logic trainer *or*
*1*  7474 dual D-type flip-flop DIP IC
*1*  7475 quad D-type flip-flop DIP IC

2   LED indicators

1   Logic probe

Several hookup wires or leads

*Procedure:*

1.  Connect the circuit shown in Figure 5-10a. Initially set the logic switches thus: $S_1$, $S_2 = 0$ and $S_3$, $S_4 = 1$.

2.  Move $S_4$ (Clr) down and up. The flip-flop should now be reset ($Q = 0$, $\overline{Q} = 1$).

3.  Move $S_1$ (*D*) up. Move $S_2$ (Clk) up and down. The flip-flop changed states when: <u>$S_2$ went up</u>, <u>$S_2$ went down</u> (circle one). Record the state of the outputs here: $Q = $ _____, $\overline{Q} = $ _____.

4.  Move $S_1$ (*D*) down. Move $S_2$ (Clk) up and down. The flip-flop changed states when: <u>$S_2$ went up</u>, <u>$S_2$ went down</u> (circle one). Record the state of the outputs here: $Q = $ _____, $\overline{Q} = $ _____.

5.  Move $S_3$ (Ps) down and up. Record the state of the outputs here: $Q = $ _____, $\overline{Q} = $ _____.

**Figure 5-10**  D-type flip-flop: (a) standard circuit; (b) D-type flip-flop wired as a T-type flip-flop; (c) quad D-type latch.

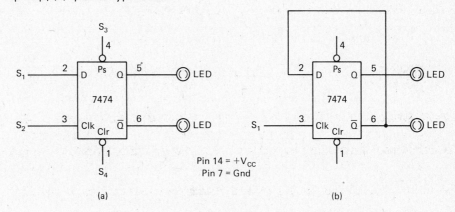

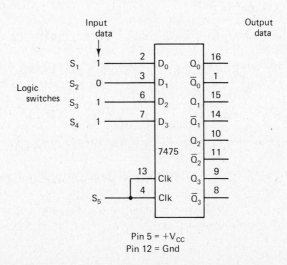

6.  Move $S_4$ (Clr) down and up. Record the state of the outputs here: $Q =$

    _____ , $\overline{Q} =$ _____ .

7.  Construct the circuit shown in Figure 5-10b to produce a T-type flip-flop. Pins 1 and 4 may not need to be connected since they will float high.

8.  Move $S_1$ (Clk) up and down several times. Notice that each time $S_1$ goes up the outputs change states.

9.  Construct the circuit as shown in Figure 5-10c. Set the logic switches thus: $S_1 = 1, S_2 = 0, S_3 = 1, S_4 = 1$, and $S_5 = 0$. There are four D-type flip-flops in this package. The first one's input is marked $D_0$ and its outputs are marked $Q_0$ and $\overline{Q}_0$. The other three flip-flops are identified in a similar manner.

10. Move $S_5$ (Clk) up and down. With a logic probe or LED indicator, test the outputs and place 0s and 1s on the figure at the respective output pins.

*Fill-In Questions:*

1.  When $D$ input $= 1$, and a clock pulse appears, the outputs will be $Q =$

    _____ and $\overline{Q} =$ _____ .

2.  When $D$ input $= 0$, and a clock pulse appears, the outputs will be $Q =$

    _____ and $\overline{Q} =$ _____ .

3.  A T-type flip-flop can be made from a D-type flip-flop by connecting

    the _____ output to the _____ input.

4.  The D-type flip-flop can function like an *R-S* flip-flop if only inputs

    _____ and _____ are used.

5.  The operation of the D-type flip-flop is that the output follows the data

    at the _____ when a clock pulse appears.

**SECTION 5-5
INSTANT REVIEW**

- A flip-flop is a bistable multivibrator that will remain in one state or the other, depending on whether it is set or reset. It has a complementary output of $Q$ and $\overline{Q}$. The flip-flop is a basic memory device and is sometimes called a latch.
- The *R-S* flip-flop has an $S$ input to turn it on and an $R$ input to turn it off.
- The clocked *R-S* flip-flop must have a pulse on the $S$ input and the clock input to turn it on and a pulse on the $R$ input and clock input to turn it off.
- The *J-K* flip-flop has a $J$ for the $S$ input and a $K$ for the $R$ input. Its nor-

mal operation is the same as that of a clocked *R-S* flip-flop. It has a preset (Ps) input which turns it on and a clear (Clr) input which turns it off. The Ps and Clr inputs override the other normal operation inputs. Bubbles at the inputs indicate that a negative-going pulse is needed for proper operation. The master/slave *J-K* flip-flop operates on the leading and trailing edges of the clock pulse.

- The T-type flip-flop has one input (*T*) and it changes state every time a pulse appears at the input.
- The D-type flip-flop operation follows the condition at the *D* input and can store a 1 for at least one clock pulse time.
- The universal *J-K* flip-flop can produce a T-type flip-flop when *J* = 1, *K* = 1, Ps = 1, and Clr = 1. If only Ps and Clr inputs are used, it will produce an *R-S* flip-flop. An inverter connected from the *J* input to the *K* input will produce a D-type flip-flop.

## SECTION 5-6
## EXERCISES II

Perform all the exercises in this section before beginning the next section.

1. With the input waveforms shown, draw the output waveforms for each of the flip-flops in Figure 5-11. Assume that all flip-flops are initially in the off state.

### Figure 5-11

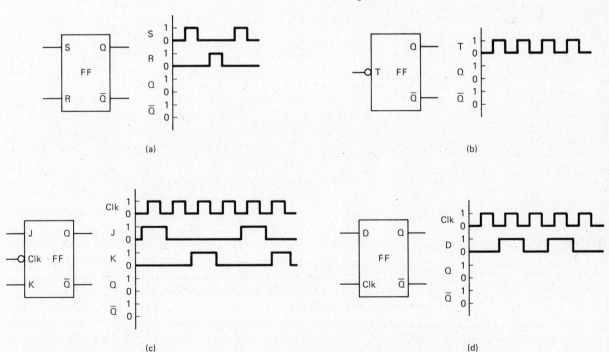

2.  Draw two different logic diagrams for converting a *J-K* flip-flop to a T-type flip-flop.

3.  Draw a logic diagram for converting a D-type flip-flop to a T-type flip-flop.

4.  Refer to Figure 5-12. With the input data given, list the condition at the outputs of the 7475 quad D latch after the enable pulse arrives.

5.  List the various flip-flop operations that can be wired with a *J-K* flip-flop by itself.

    1. _____ flip-flop

    2. _____ flip-flop

    3. _____ flip-flop

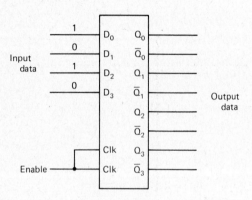

**Figure 5-12**

6.  With the input waveforms given, draw the output waveforms for the *J-K* flip-flop shown in Figure 5-13.

**Figure 5-13**

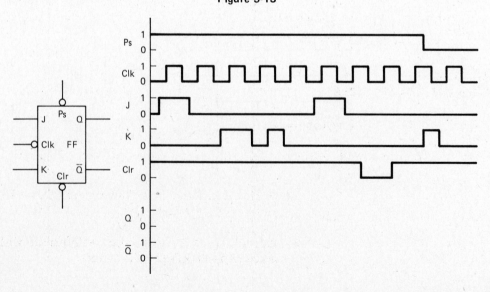

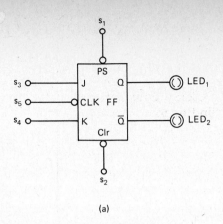

(a)

| Condition | PS $s_1$ | Clr $s_2$ | J $s_3$ | K $s_4$ | CLK $s_5$ | Q | $\overline{Q}$ | Comments |
|-----------|----------|-----------|---------|---------|-----------|---|----------------|----------|
| Normally off | 1 | 1 | 0 | 0 | ↓↑ | | | No action |
| Normally on | 1 | 1 | 1 | 0 | ↓↑ | | | FF turns on |
| Normally reset | 1 | 1 | 0 | 1 | ↓↑ | | | FF turns off |
| Preset | 0 | 1 | 0 | 1 | ↓↑ | | | FF will not turn off |
| Clear | 1 | 0 | 1 | 0 | ↓↑ | | | FF will not turn on |
| Toggle | 1 | 1 | 1 | 1 | ↓↑ | | | FF will turn on |
| Toggle | 1 | 1 | 1 | 1 | ↓↑ | | | FF will turn off |

(b)

**Figure 5-14** Troubleshooting the *J-K* flip-flop: (a) logic diagram; (b) testing chart.

## SECTION 5-7
## TROUBLESHOOTING APPLICATION: TESTING THE *J-K* FLIP-FLOP

Finding a malfunctioning flip-flop means to test it for its normal and full operation. The *J-K* flip-flop is used here since it is a universal flip-flop.

Connect the *J-K* flip-flop to the logic switches and LEDs as shown in Figure 5-14a. Set the Ps, Clr, *K*, and Clk switches to a 1. Set the *J* switch to a 0. Momentarily operate the Clr switch ($S_2$). The flip-flop should now be in the normally off state ($Q = 0$, $\overline{Q} = 1$). Set the switches as shown in the second row of the testing chart of Figure 5-14b. Momentarily operate the Clk switch ($S_5$), indicated by ↑↓. The flip-flop should not change states. Use this method for the remaining rows, setting the switches as indicated, and record a 1 when the outputs are on and a 0 when they are off.

## SECTION 5-8
## SELF-CHECKING QUIZZES

### 5-8a  FLIP-FLOPS: TRUE-FALSE QUIZ

Place a T for true or an F for false to the left of each statement.

_____  1. A T-type flip-flop will change states with a 1 pulse to the T input and a 0 pulse to the clock.

_____  2. A negated clock *J-K* flip-flop will turn on with a 1 pulse to the *K* input and a 0 pulse to the clock.

153

_____ 3. Preset and clear inputs can override the normal inputs of a _J-K_ flip-flop.

_____ 4. A D-type flip-flop can delay the transfer of data from input to output by one clock pulse.

_____ 5. An _R-S_ flip-flop has only two inputs.

_____ 6. Another name for a flip-flop is "latch."

_____ 7. The _J-K_ flip-flop and the D-type flip-flop are asynchronous in operation.

_____ 8. When the $Q$ output of a flip-flop is 1, its $\overline{Q}$ output will be 0.

_____ 9. A master/slave _J-K_ flip-flop operates on the leading and trailing edges of the clock pulse.

_____ 10. A flip-flop is considered on when its $Q$ output = 1 and $\overline{Q}$ output = 0.

## 5-8b  FLIP-FLOPS: MULTIPLE-CHOICE QUIZ

Circle the correct answer for each question.

1. The operation of a flip-flop is known as:
   a. monostable
   b. bistable
   c. astable
   d. quasi-stable

2. The _J-K_ flip-flop will operate as a T-type flip-flop with each input clock pulse when inputs;
   a. $J = 1, K = 1, Ps = 1, Clr = 1$
   b. $J = 1, K = 0, Ps = 1, Clr = 1$
   c. $J = 0, K = 0, Ps = 1, Clr = 1$
   d. $J = 1, K = 1, Ps = 0, Clr = 1$

3. The _J-K_ flip-flop can be operated as an _R-S_ flip-flop by using only inputs:
   a. _J_, _K_, and Clr
   b. _J_, _K_, and Clk
   c. Ps and Clr
   d. Ps, Clr, and Clk

4. With normal operation a negated clock _J-K_ flip-flop will turn on when:
   a. $J = 0, K = 1, Ps = 1, Clr = 1, Clk = \downarrow$
   b. $J = 1, K = 0, Ps = 1, Clr = 1, Clk = \downarrow$
   c. $J = 1, K = 0, Ps = 1, Clr = 0, Clk = \downarrow$
   d. $J = 0, K = 0, Ps = 0, Clr = 0, Clk = \downarrow$

5. The flip-flip that automatically resets on the next clock pulse when the data on its input line go to 0 is the:
   a. _J-K_
   b. T-type
   c. D-type
   d. _R-S_

6. If a T-type flip-flop is on, after three input pulses its outputs are:
   a. $Q = 0, \overline{Q} = 1$
   b. $Q = 1, \overline{Q} = 0$
   c. $Q = 1, \overline{Q} = 1$
   d. $Q = 0, \overline{Q} = 0$

7. When a flip-flop is controlled with clock pulses, its operation is called:

   a. asynchronous

   b. synchronous

   c. bistable

   d. astable

8. A T-type flip-flop can be constructed from a:

   a. *J-K* flip-flop

   b. *R-S* flip-flop

   c. D-type flip-flop

   d. answers a and b are correct

   e. answers a and c are correct

9. Referring to Figure 5-15, with $S_1$ in position $A$ as shown:

   a. $Q = 0, \overline{Q} = 1$

   b. $Q = 1, \overline{Q} = 1$

   c. $Q = 0, \overline{Q} = 0$

   d. $Q = 1, \overline{Q} = 0$

10. Referring to Figure 5-15, when $S_1$ is moved to position $B$:

   a. $Q = 0, \overline{Q} = 1$

   b. $Q = 1, \overline{Q} = 1$

   c. $Q = 0, \overline{Q} = 0$

   d. $Q = 1, \overline{Q} = 0$

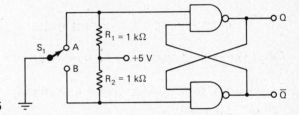

**Figure 5-15**

## ANSWERS TO EXPERIMENTS AND QUIZZES FOR UNIT 5

*Experiment 1.*

   **(1)** forbidden    **(2)** each or every    **(3)** synchronous    **(4)** high, low

*Experiment 2.*

   **(1)** Ps, Clr    **(2)** on    **(3)** off    **(4)** T-type    **(5)** 0, 1    **(6)** 1, 0

*Experiment 3.*

   **(1)** 1, 0    **(2)** 0, 1    **(3)** $\overline{Q}, D$    **(4)** Ps, Clr    **(5)** *D* input

*True-False:*

   **(1)** F    **(2)** F    **(3)** T    **(4)** T    **(5)** T    **(6)** T    **(7)** F    **(8)** T    **(9)** T
   **(10)** T

*Multiple-Choice:*

   **(1)** b    **(2)** a    **(3)** c    **(4)** b    **(5)** c    **(6)** a    **(7)** b    **(8)** e    **(9)** d
   **(10)** a

# Unit 6

## Binary Registers

A flip-flop is a basic memory cell or unit. When a group of flip-flops are connected together (one after another), they form what is called a *binary register*. The function of a binary register is to provide temporary data storage. The number of flip-flops in a register determines the amount of data per unit, referred to as the *data word* or *binary word*. A register with a word length of 4 bits contains four flip-flops. Terms are given to registers with specific word lengths for ease of understanding. A word length of 4 bits is called a *nibble* and a word length of 8 bits is called a *byte*.

Binary registers are very important and very useful in digital circuits and computers for performing various operations. Various registers are designed and constructed to perform specific tasks or operations. Some registers may perform some tasks, while others perform different tasks. Regardless of the versatility of a register, it nearly always needs external or extra control circuitry to perform other desired operations. This unit will show the following operations involving registers:

1. Serial input data loading
2. Parallel input data loading
3. Serial output data
4. Parallel output data
5. Clearing data
6. Shifting data right
7. Shifting data left

8. Rotating data
9. Complementing data
10. Serial-to-parallel data conversion
11. Parallel-to-serial data conversion
12. Multiplication by factors of 2
13. Division by factors of 2

## 6-1b  THE BASIC 4-BIT WORD BINARY REGISTER

The basic 4-bit word binary register is shown in Figure 6-1a. Four *J-K* flip-flops (FFs) are connected together. The $Q$ output of one FF is connected to the $J$ input of the following FF and the $\overline{Q}$ output is connected to the $K$ input. All the clock inputs of the FFs are connected together and brought to a common point called the clock (Clk). All of the clear inputs are connected together and brought to a common point called clear (Clr). The data that are being stored in a register can be represented by LEDs connected to the $Q$ outputs. Since we normally read right to left, the flip-flops are labeled $FF_A$, $FF_B$, $FF_C$, and $FF_D$; however, for positional notation used in mathematics the weighting of the LEDs are reversed (i.e., $FF_A = 2^0$ or 1, $FF_B = 2^1$ or 2, $FF_C = 2^2$ or 4, and $FF_D = 2^3$ or 8). The letters LSB stand

**Figure 6-1**  Four-bit word binary register; (a) logic diagram; (b) register table.

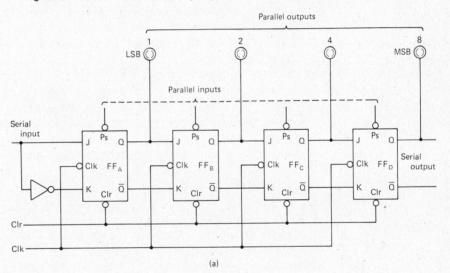

(a)

| Serial input | Clock pulses | | FF_A FF_B FF_C FF_D | | | |
|---|---|---|---|---|---|---|
| 1 | 1 | LSB | 1 | 0 | 0 | 0 | MSB | Register loaded with a 1 |
| 0 | 2 | LSB | 0 | 1 | 0 | 0 | MSB | 1st shift |
| 0 | 3 | LSB | 0 | 0 | 1 | 0 | MSB | 2nd shift |
| 0 | 4 | LSB | 0 | 0 | 0 | 1 | MSB | 3rd shift |
| 0 | 5 | LSB | 0 | 0 | 0 | 0 | MSB | 4th shift, data lost |

(b)

for *least significant bit* (or lowest order) and MSB stands for *most significant bit* (or highest order) of a number. This may or may not be the case in other circuits, and you will have to give attention to the particular circuit you are working on within the various equipment. If it is desired to have the normal order of positional notation, the LEDs can be cross-connected to the flip-flops, such as LED 1 to $FF_D$, LED 2 to $FF_C$, LED 4 to $FF_B$, and LED 8 to $FF_A$. Now MSB would be to the far left and LSB would be to the far right.

### 6-1b.1    Serial Data Input

The serial data input is to the $J$ input of $FF_A$. When this data input is 1 and a clock pulse appears, $FF_A$ turns on. The inverter connected from $J$ input to the $K$ input forms a D-type FF and allows $FF_A$ to turn off when the data line is 0 and a clock pulse appears. This operation is called *serial data loading*.

### 6-1b.2    Shift Right Operation

The *shift right operation* can be understood by the register table shown in Figure 6-1b. Initially, the serial input line is 1 and when the first clock pulse appears, the 1 is loaded into $FF_A$. At this time the $Q$ output of $FF_A$ is a 1, which is applied to the $J$ input of $FF_B$. The $\overline{Q}$ output of $FF_A$ is a 0, which is applied to the $K$ input of $FF_B$. A *steering on* action is now set up for $FF_B$.

To show basically how a single bit of datum can be shifted through a register, the serial input line goes to 0. A *steering off* action is now set up for $FF_A$. When the second clock pulse appears, $FF_B$ turns on while $FF_A$ turns off. Now the $Q$ output = 0 and the $\overline{Q}$ output = 1 of $FF_A$, which steers $FF_B$ to turn off, while $FF_B$ steers $FF_C$ to turn on when the third clock pulse appears. This procedure is repeated as subsequent clock pulses occur; that is, the fourth clock pulse shifts the 1 to $FF_D$ and the fifth clock pulse shifts the 1 out of the register and the data are lost. Notice that after the register is loaded, it takes four clock pulses to shift the data out of the register. For an 8-bit register it would take eight clock pulses to clear the register to zero.

### 6-1b.3    Serial Output Data

When data are shifted right, the $Q$ and $\overline{Q}$ outputs of $FF_D$ provide serial output data for another register or circuit that may be connected to these points. If another register is connected to $FF_D$, its Clk line must be connected to or synchronized with the Clk line of the first register in order to accept the data.

### 6-1b.4    Parallel Output Data

Once the register is loaded, its data are available for parallel use or operation at the $Q$ outputs of the FFs in the register.

### 6-1b.5    Serial-to-Parallel Data Conversion

Data can enter a register serially, perhaps from an external source on two lines, the serial input and the ground line (usually not shown). The data can then be transferred in a parallel manner, from the $Q$ outputs of the FFs to other internal circuits. This is a basic operation of *serial-to-parallel data conversion*.

### 6-1b.6  Parallel Input Data

Data can enter a register in a parallel manner through the preset (Ps) inputs of the FFs. In this case the data would have to be complemented since a low or negative-going pulse turns on the FFs. For example, if the binary number LSB 1 0 1 0 MSB was to be entered, the data present at the Ps inputs would have to be LSB 0 1 0 1 MSB.

### 6-1b.7  Parallel-to-Serial Data Conversion

Once the data are loaded in parallel into a register, they can now be shifted to the right out of $FF_D$ to another circuit or pair of lines in a serial manner. This is a basic operation of *parallel-to-serial data conversion.*

### 6-1b.8  Clearing Data from a Register

The data in a register may no longer be needed. Instead of taking time to shift them out of the register, a low or negative-going pulse to the Clr line will reset all the FFs in the register. Now, all the $Q$ outputs will be 0. This operation is referred to as *clearing a register.*

## 6-1c  SHIFT RIGHT/SHIFT LEFT REGISTER

The 7495 IC contains a 4-bit register with additional logic gates as shown in Figure 6-2a. These additional circuits enable the 7495 to shift data left or right. However, for a shift left operation the $Q$ outputs of the FFs must be connected back to the inputs of the preceding FFs. A 7432 quad two-input OR gate IC is used to accomplish this while also allowing data to be parallel loaded into the register.

The 7495 IC has a separate serial data input, four parallel data inputs, four parallel data outputs with $Q_D$ acting as the serial data output, a mode control (mode ctrl) input, one clock input for shifting right (Clk 1 SHR), and one clock input for shifting left (Clk 2 SHL). In some applications the two clock inputs are connected together. To shift data to the right, the mode ctrl input must be 0 and the clock pulses present at the SHR clock input. There are also inputs for $+V_{CC}$ and Gnd.

### 6-1c.1  Shift Left Operation

To *shift left*, the mode ctrl input must be a 1 and clock pulses present at the SHL clock input. To show one example of a SHL operation, assume $Q_C$ to be a 1, which is applied to one input of $OR_2$. This makes one input of $AN_4$ a 1. If the mode ctrl input is 1, then the output of $I_1$ is 0 and the output of $I_2$ is a 1, which is applied to the other input of $AN_4$. The output of $AN_4$ is a 1, which causes the output of $NR_2$ to be 0; therefore, the $R$ input of $FF_B$ is 0 and its $S$ input is 1. When a clock pulse arrives, $FF_B$ will turn on. Similarly, a 0 applied from a $Q$ output to the input of a preceding FF will turn it off.

### 6-1c.2  Multiplication by Factors of 2

The register table of Figure 6-2b shows how a binary 3 is shifted on a SHR operation. Notice that after the first shift the register now contains a binary

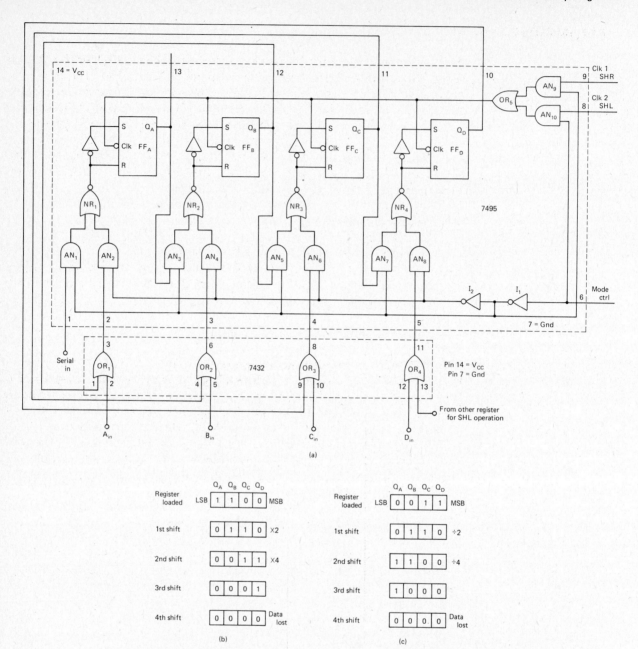

**Figure 6-2**  Shift right/shift left register: (a) 7495 IC logic diagram; (b) SHR register table; (s) SHL register table.

6, and after the second shift it contains a binary 12. Shifting right then produces multiplication by factors of 2.

### 6-1c.3  Division by Factors of 2

The register table of Figure 6-2c shows how a binary 12 is shifted on a SHL operation. After the first shift the register contains a binary 6, and after the second shift the register contains a binary 3. Shifting left then produces division by factors of 2.

## 6-1d  ROTATING DATA

Sometimes it is necessary to serially shift data out of a register to other circuits, but also retain the original data in the register. This can be accomplished by rotating or recirculating the data out of the serial output and feeding it back into the serial input as illustrated in Figure 6-3a. In this case the data leave the serial output and pass through an exclusive-OR gate before entering the serial input, enabling the original data or their complement to be put back into the register. (Review the operation of an exclusive-OR gate in Section 2-1f.) Since this is a 4-bit word register, it takes four clock pulses to complete the rotate operation.

### 6-1d.1  Normal Data Rotate

In order for the original data to be rotated back into the register, the control input of $EOR_1$ must be a 0. As the clock pulses appear, the data are shifted out of the register and back into their serial input simultaneously, as shown by the normal rotate register table of Figure 6-3b.

### 6-1d.2  Complement Data Rotate

If it is desired to complement the data in the register, the control input of $EOR_1$ must be a 1. When the clock pulses appear, the data shifted out of $EOR_1$ back to the serial input are the complement of the original data, as shown by the complement rotate register table of Figure 6-3c.

**Figure 6-3**  Rotate or recirculate data operation: (a) logic diagram; (b) normal rotate register table; (c) complement rotate register table.

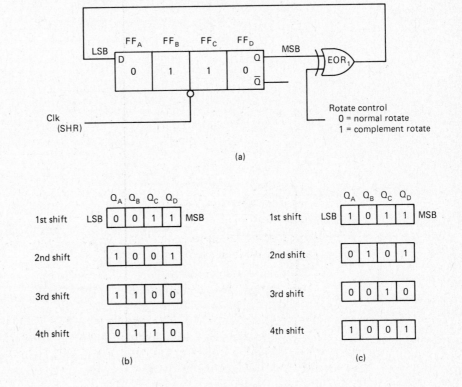

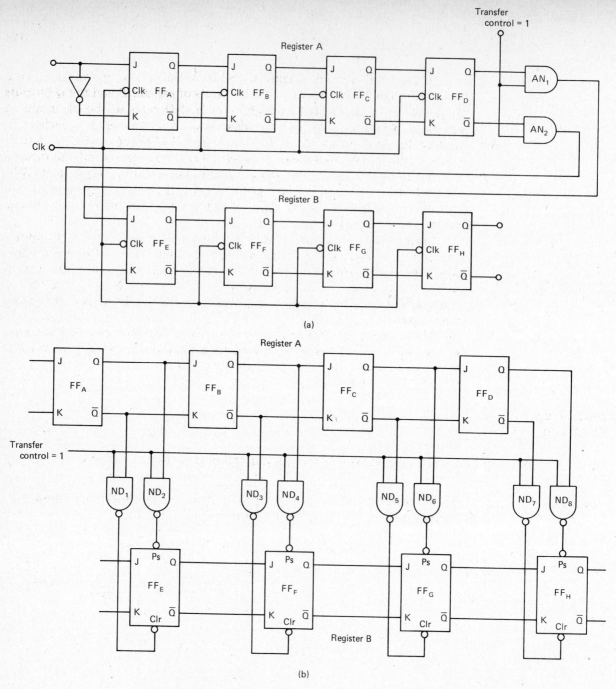

**Figure 6-4** Register data transfer: (a) serial mode; (b) parallel mode.

## 6-1e REGISTER DATA TRANSFER

In digital operations and computers it is necessary to transfer data from one register to another register. The two methods of accomplishing this data transfer is by the serial mode and the parallel mode, as shown in Figure 6-4.

In the serial mode (Figure 6-4a) register $A$ is connected serially to register $B$ through two transfer control AND gates. The $Q$ output of $FF_D$ goes through $AN_1$ to the $J$ input of $FF_E$ and the $\overline{Q}$ output of $FF_D$ goes through $AN_2$ to the $K$ input of $FF_E$. When it is desired to transfer data from register $A$ to register $B$, the transfer control line of the AND gates must be 1. When the clock pulses arrive, data are shifted from register $A$ into register $B$. With this operation the data in register are lost unless a rotate operation is used simultaneously to place the data back into it. Notice that both registers must use the same clock pulses or be in synchronization.

In the parallel mode (Figure 6-4b) the $Q$ outputs of register $A$ are connected through NAND gates to the Ps inputs of register $B$ and the $\overline{Q}$ outputs of register $A$ are connected through other NAND gates to the Clr inputs of register $B$. When it is desired to transfer data from register $A$ to register $B$, the transfer control line must go to a 1. The NAND gates with 1s at their inputs from register $A$ will then produce a 0 at their outputs and will turn on or turn off the FFs in register $B$ accordingly. For example, if $FF_A$ is on, $ND_2$ will produce a 0 to set $FF_E$, and similarly if $FF_B$ is off, $ND_3$ will produce a 0 to reset (or keep off) $FF_F$. This type of transfer does not require clock pulses and the data are retained in register $A$.

Notice that the advantage of serial mode data transfer is that it requires fewer control gates, but its disadvantages are that clock pulses are needed, more time is required, and the data are lost in register $A$. The advantages of parallel mode data transfer is that clock pulses are not needed, less time is required, and the data are not lost in register $A$. Its disadvantage is that more control gates are required.

These two methods of data transfer between registers are used in digital systems and the compromise between their advantages and disadvantages is known as the *trade-off factor* in designing circuits.

## SECTION 6-2
## EXERCISES I

Perform all the exercises in this section before beginning the next section.

1. Draw a 4-bit word register using *J-K* flip-flops (refer to Figure 6-1a).

2. Indicate how many flip-flops are used in the following registers:

   a. 4-bit word = _____ FFs

   b. 8-bit word = _____ FFs

   c. 16-bit word = _____ FFs

3. Indicate the contents of the register after each shift for a shift right operation with the data given (refer to Figure 6-1b).

| Loaded | LSB | 1 | 1 | 0 | 1 | MSB |
|--------|-----|---|---|---|---|-----|
| 1st shift | | ___ | ___ | ___ | ___ | |
| 2nd shift | | ___ | ___ | ___ | ___ | |
| 3rd shift | | ___ | ___ | ___ | ___ | |
| 4th shift | | ___ | ___ | ___ | ___ | |

4. Indicate the contents of the register after each shift for a shift left operation with the data given (refer to Figure 6-2c).

| Loaded | LSB | 1 | 0 | 1 | 0 | MSB |
|--------|-----|---|---|---|---|-----|
| 1st shift | | ___ | ___ | ___ | ___ | |
| 2nd shift | | ___ | ___ | ___ | ___ | |
| 3rd shift | | ___ | ___ | ___ | ___ | |
| 4th shift | | ___ | ___ | ___ | ___ | |

5.  Indicate the contents of the register after each shift for a normal rotate operation with the data given (refer to Figure 6-3b).

| Loaded | LSB | 1 | 0 | 1 | 1 | MSB |
|--------|-----|---|---|---|---|-----|

1st shift  ___ ___ ___ ___

2nd shift  ___ ___ ___ ___

3rd shift  ___ ___ ___ ___

4th shift  ___ ___ ___ ___

6.  Indicate the contents of the register after each shift for a complement rotate operation with the data given (refer to Figure 6-3c).

| Loaded | LSB | 1 | 0 | 1 | 1 | MSB |
|--------|-----|---|---|---|---|-----|

1st shift  ___ ___ ___ ___

2nd shift  ___ ___ ___ ___

3rd shift  ___ ___ ___ ___

4th shift  ___ ___ ___ ___

7.  Draw the logic diagram showing serial mode data transfer between two registers.

8.  Draw the logic diagram showing parallel mode data transfer between two registers.

9.  List the advantages and disadvantages of each type of data transfer mode between registers.

*Serial Transfer Mode*

Advantage       1._____

Disadvantages   1._____

                2._____

                3._____

*Parallel Transfer Mode*

Advantages      1._____

                2._____

                3._____

Disadvantage    1._____

10. A binary register is capable of multiplying and dividing evenly by factors of 2. Indicate which type of shift will accomplish each operation as given in this text.

   a. Multiplication = shift _____    b. Division = shift _____


**SECTION 6-3**
**DEFINITION EXERCISES**

Give a brief description of each of the following terms.

1. Binary word _____

_____

_____

2. Nibble _____

_____

_____

3. Byte _____

_____

_____

4. Register _____

_____

_____

5. Serial input data operation _____

_____

_____

6. Parallel input data operation _____

_____

_____

7. Serial output data operation _____

_____

_____

8.  Parallel output data operation _____

_____

_____

9.  Shift right operation _____

_____

_____

10.  Shift left operation _____

_____

_____

11.  Complement of 0 _____

_____

_____

12.  Complement of 1 _____

_____

_____

13.  Complement operation _____

_____

_____

14.  Rotate data operation _____

_____

_____

15.  Recirculate data operation _____

_____

_____

16.  Clear operation _____

_____

_____

17.  Data conversion: serial to parallel _____

_____

_____

18.  Data conversion: parallel to serial _____

_____

_____

19.  Serial mode data transfer _____

_____

_____

20.  Parallel mode data transfer _____

_____

_____

21.  Trade-off factor _____

_____

_____

22.  LSB _____

_____

_____

23.  MSB _____

_____

_____

**SECTION 6-4**
**EXPERIMENTS**

**EXPERIMENT 1.   BASIC 4-BIT SHIFT REGISTER**

*Objective:*

To demonstrate the operations of a basic shift right 4-bit register constructed from separate flip-flops.

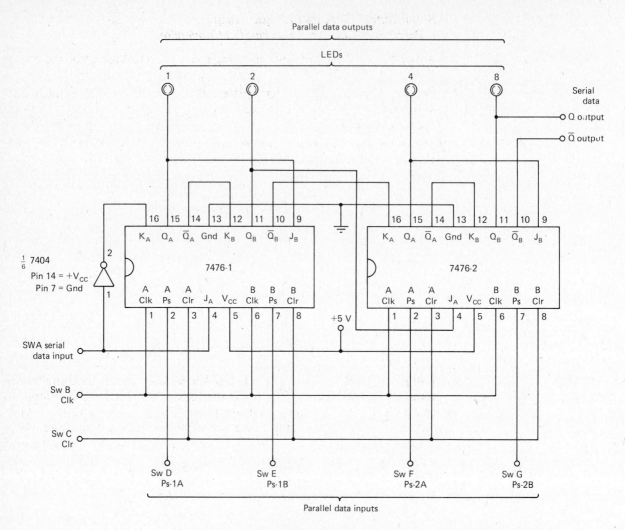

**Figure 6-5** Four-bit word register using 7476 ICs.

*Introduction:*

The operations capable with this register are: serial input, serial output, parallel input, parallel output, shift right, clear, serial-to-parallel conversion, and parallel-to-serial conversion. The inverter at the inputs of $FF_A$ make this a D-type FF. Refer to Figure 6-1 for a logic analysis of the circuit.

*Materials Needed:*

*1* Digital logic trainer *or*
*2* 7476 dual *J-K* master/slave flip-flop DIP ICs
*1* 7404 hex inverter DIP IC
*4* LED indicators
  Several hookup wires or leads

*Procedure:*

1. Construct the circuit shown in Figure 6-5.
2. Place the switches in the following positions: Sw $A$ = low, Sw $B$ = high, Sw $C$ = high, and Sw $D$ through Sw $G$ = high.

*Serial Input, Shift Right, Serial Output,*
  *Parallel Output, Serial-to-Parallel Conversion*

3. Move Sw $C$ (Clr) low and then high. This resets the entire register to zero.

4. Set Sw $A$ (serial input) high. Move Sw $B$ (Clk) low and then high. LED 1 should turn on, indicating that $FF_A$ is on. Move Sw $A$ low. The register is now loaded with a 1.

5. Move Sw $B$ (Clk) down and up as indicated ↓↑ and record the conditions at the parallel outputs (0 = off, 1 = on) after each shift here:

| Clk | Shift | $Q_A$ | $Q_B$ | $Q_C$ | $Q_D$ |
|-----|-------|-------|-------|-------|-------|
| ↓↑ | 1. | | | | |
| ↓↑ | 2. | | | | |
| ↓↑ | 3. | | | | |
| ↓↑ | 4. | | | | |

6. Using the same procedure as in step 4, load the register with the binary number LSB 1011 MSB. Notice that with the serial input operation the data must be shifted into the register. When the data word contains a 0, Sw $A$ must be low as the data are shifted in. The data in the register are available at outputs $Q_A$, $Q_B$, $Q_C$, and $Q_D$ for a parallel-out operation and shows serial-to-parallel conversion.

7. Similar to step 5, shift the data to the right using Sw $B$ (Clk) and record the conditions of the outputs after each shift here:

| Clk | Shift | $Q_A$ | $Q_B$ | $Q_C$ | $Q_D$ |
|-----|-------|-------|-------|-------|-------|
| ↓↑ | 1. | | | | |
| ↓↑ | 2. | | | | |
| ↓↑ | 3. | | | | |
| ↓↑ | 4. | | | | |

*Parallel Input, Clearing, Shift Right, Parallel-to-Serial Conversion*

8. Move Sw $D$ and Sw $F$ low and then high. LEDs 1 and 4 should be on, indicating that the register contains the data word LSB 1010 MSB.

9. Move Sw $C$ (Clr) low and then high. The LEDs should go out and the register is cleared or reset.

10. Again, move Sw $D$ and Sw $F$ low and then high. The register contains the data word LSB 1010 MSB.

11. Again, shift the data right using Sw $B$ (Clk) and record the conditions of the outputs after each shift here:

| Clk | Shift | $Q_A$ | $Q_B$ | $Q_C$ | $Q_D$ |
|-----|-------|-------|-------|-------|-------|
| ↓↑ | 1. | | | | |
| ↓↑ | 2. | | | | |
| ↓↑ | 3. | | | | |
| ↓↑ | 4. | | | | |

These last two steps show an example of parallel-to-serial conversion. The data could have been shifted to another register or piece of equipment if a wire was used at output $Q_D$.

*Fill-In Questions:*

1. With a serial input operation the data must be _____ into the register.

2. In a parallel output operation data can be taken from outputs _____, _____, _____, and _____.

3. For this type of register, during a shifting operation the $Q$ of one FF steers the _____ input of the following FF, while the $\overline{Q}$ steers the _____ input.

4. The necessary factor for shifting is _____ pulses.

5. When the Clr line goes low, the register _____ to zero.

6. A parallel input operation uses the _____ inputs to the FFs directly.

7. The two modes of data conversion with this register are _____ and _____.

## EXPERIMENT 2.  SHIFT RIGHT/SHIFT LEFT REGISTER

*Objective:*

To show how the 7495 shift right/shift left register IC can be wired for proper operation.

*Introduction:*

The 7495 can be wired directly from the outputs to the inputs to accomplish both directions of shifting, however, the register can be loaded only by the serial input. With the addition of OR gates in the form of a 7432 IC at the parallel inputs, the register can be loaded either serially or directly with the parallel input operation. Refer to Figure 6-2 for a logic analysis of the circuit.

*Materials Needed:*

*1* Digital logic trainer *or*

*1* 7495 4-bit shift right/shift left register DIP IC

*1* 7432 quad two-input OR gate DIP IC

*4* LED indicators

Several hookup wires or leads

*Procedure:*

1. Connect the circuit as shown in Figure 6-6.
2. Set all switches to 0 or low.

**Figure 6-6**  7495 IC modified for shift right/shift left operation.

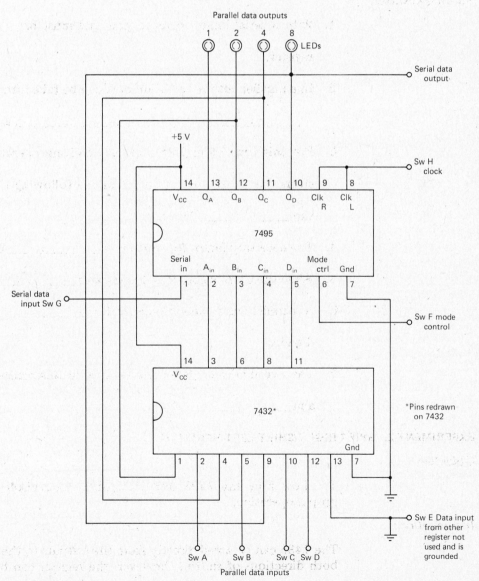

*Shift Right Operation*

3.  Set Sw *A* and Sw *C* to a high or 1 and Sw *F* to a low or 0.
4.  Move Sw *H* (Clk) high and then low. LEDs 1 and 4 should turn on.
5.  Return Sw *A*, Sw *C*, and Sw *F* to a 0 or low. The register now contains the binary number LSB 1010 MSB.
6.  Perform a shift right (SHR) operation by moving Sw *H* (Clk) high and then low. After each shift record the conditions of the outputs here:

| Clk | Shift | $Q_A$ | $Q_B$ | $Q_C$ | $Q_D$ |
|-----|-------|-------|-------|-------|-------|
| ↑↓  | 1.    |       |       |       |       |
| ↑↓  | 2.    |       |       |       |       |
| ↑↓  | 3.    |       |       |       |       |
| ↑↓  | 4.    |       |       |       |       |

Notice that the register initially contained a binary 5 and after the first shift right it contained a binary 10, a multiplication factor of 2. Since the register has only 4 bits, subsequent shifts resulted in the data being lost.

*Shift Left Operation*

7.  Set Sw *B*, Sw *D*, and Sw *F* to a high or 1.
8.  Move Sw *H* (Clk) high and then low. LEDs 2 and 8 should turn on.
9.  Return Sw *B* and Sw *D* to a 0 or low. The register now contains the binary number LSB 0101 MSB.
10. Leave Sw *F* (mode control) high.
11. Perform a shift left (SHL) operation by moving Sw *H* (Clk) high and then low. After each shift left, record the conditions of the outputs here:

| Clk | Shift | $Q_A$ | $Q_B$ | $Q_C$ | $Q_D$ |
|-----|-------|-------|-------|-------|-------|
| ↑↓  | 1.    |       |       |       |       |
| ↑↓  | 2.    |       |       |       |       |
| ↑↓  | 3.    |       |       |       |       |
| ↑↓  | 4.    |       |       |       |       |

Notice that initially, the register contained a binary 10 and after the first shift left it contained a binary 5, a division factor of 2. Since the register has only 4 bits, subsequent shifts resulted in the data being lost.

*Fill-In Questions:*

1. To parallel load this register, the mode control input must be _____

   _____.

2. To perform a SHR operation, the mode control input must be _____

   _____.

3. A SHR operation can provide the mathematical function of _____

   _____ by factors of 2.

4. To perform a SHL operation, the mode control input must be _____

   _____.

5. A SHL operation can provide the mathematical function of _____

   _____ by factors of 2.

6. Data are quickly lost in a shifting operation because this register only

   contains _____ bits.

7. It takes _____ shifts to clear this register to zero.

## EXPERIMENT 3.  THE 7495 AS A MORE VERSATILE REGISTER

*Objective:*

To demonstrate how the 7495 shift register IC with additional logic circuits can be used for a complement operation while retaining other desired features.

*Introduction:*

This circuit can accomplish the following register operations: serial input, serial output, SHR, SHL, complementing, parallel input, and parallel output.

*Materials Needed:*

*1* Digital logic trainer *or*
*1* 7495 4-bit shift right/shift left register DIP IC
*1* 7432 quad two-input OR gate DIP IC
*2* 7451 dual two-wide two-input AND-OR-Inverter gate DIP ICs
*1* 7404 hex inverter
*4* LED indicators
   Several hookup wires or leads

*Procedure:*

Since experience was gained in Experiment 2 working with the 7495 IC, the procedures will be given in tabular form to simplify operations.

1. Construct the circuit shown in Figure 6-7.

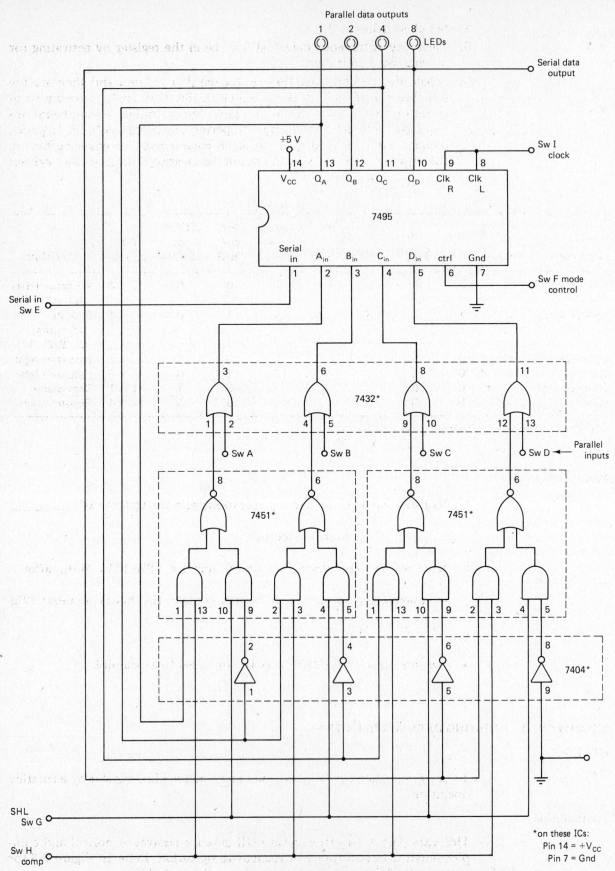

**Figure 6-7** 7495 IC modified for various operations.

2. Set all switches to 0.

3. Clear any extraneous data that may be in the register by activating (or moving) Sw I four times.

4. Select the desired operation in the far left column and then set the switches accordingly. Activate the clock (Sw I) the indicated number of times (example: ↑↓ = 4) to complete the operation. Some operations will have to be grouped together to perform an overall task, for example, *parallel input* for loading the register, *complement* for changing data in the register, *SHR* or *SHL* operation for clearing data from the register.

| Operation | Sw A | Sw B | Sw C | Sw D | Serial in Sw E | Mode ctrl Sw F | SHL ctrl Sw G | COMP ctrl Sw H | Clk Sw I | Comments |
|---|---|---|---|---|---|---|---|---|---|---|
| Serial input | 0 | 0 | 0 | 0 | X[a] | 0 | 0 | 0 | ↑↓ = 4 | Set desired data for each shift |
| Parallel input | X | X | X | X | 0 | 1 | 1 | 0 | ↑↓ = 1 | Set data at inputs A, B, C, D |
| Shift right | 0 | 0 | 0 | 0 | 0 | 0 | 0 | 0 | ↑↓ = 4 | Data shift right |
| Shift left | 0 | 0 | 0 | 0 | 0 | 1 | 1 | 0 | ↑↓ = 4 | Data shift left |
| Complement | 0 | 0 | 0 | 0 | 0 | 1 | 0 | 1 | ↑↓ = 1 | Data change |
| Clear | 0 | 0 | 0 | 0 | 0 | 0 | 0 | 0 | ↑↓ = 4 | Register resets |

[a]The X indicates the desired data 0 or 1, to be entered into the register.

*Fill-In Questions:*

1. Registers can perform various operations with the addition of _____ _____ control circuits.

2. If a register contained the binary number LSB 1011 MSB, after a complementing operation it would contain the binary number LSB __ __ __ __ MSB.

3. Various register operations may be combined to accomplish an _____ _____ _____.

**EXPERIMENT 4.  ROTATING DATA IN A REGISTER**

*Objective:*

To show how data can be saved and complemented in a register by a rotating operation.

*Introduction:*

This experiment uses an exclusive-OR gate for control of normal and complemented data in a rotate or recirculate operation. Refer to Figure 6-3 for an analysis of the contents of the register after each shift.

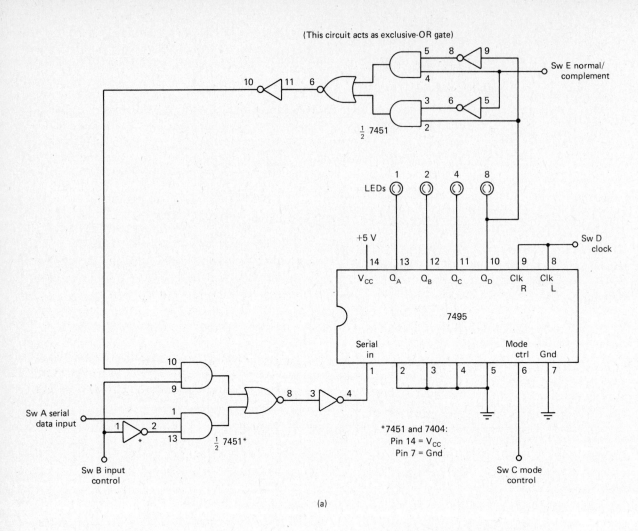

(This circuit acts as exclusive-OR gate)

*7451 and 7404:
Pin 14 = V_CC
Pin 7 = Gnd

(a)

| Operation | Sw A data | Sw B input ctrl | Sw C mode ctrl | Sw D Clk | Sw E rotate N/C |
|---|---|---|---|---|---|
| Serial enter (load register) | 1 | 0 | 0 | ↑↓ | 0 |
| Normal rotate | 0 | 1 | 0 | ↑↓ | 0 |
| Complement rotate | 0 | 1 | 0 | ↑↓ | 1 |
| Clear | 0 | 0 | 0 | ↑↓ | 0 |

(b)     *Depends if data is 1 or 0

| Action | $Q_A$ | $Q_B$ | $Q_C$ | $Q_D$ | |
|---|---|---|---|---|---|
| Register loaded | 1 | 1 | 0 | 1 | |
| 1st shift | | | | | Normal rotate |
| 2nd shift | | | | | |
| 3rd shift | | | | | |
| 4th shift | | | | | |
| | | | | | |
| 1st shift | | | | | Complement rotate |
| 2nd shift | | | | | |
| 3rd shift | | | | | |
| 4th shift | | | | | |

(c)

**Figure 6-8** Rotating data in a register: (a) logic diagram; (b) operation/select switch table; (c) register table.

*Materials Needed:*

1   Digital logic trainer *or*
1   7495 4-bit shift right/shift left register DIP IC
1   7451 dual two-wide two-input AND-OR-Inverter gate DIP IC
1   7404 hex inverter DIP IC
    Several hookup wires or leads

*Procedure:*

1.  Connect the circuit as shown in Figure 6-8a.

    Using the operation/select switch table of Figure 6-8b:

2.  Enter the binary number LSB 1101 MSB into the register.
3.  Set the switches for a normal rotate operation and record the conditions of the outputs of the register after each shift in the upper portion of the register table shown in Figure 6-8c. Are the rotated data the same as the

    original data? _____.

4.  Set the switches for a complement rotate operation and record the conditions of the outputs of the register after each shift in the lower portion of the register table shown in Figure 6-8c. Is the complement correct?

    _____.

5.  Set the switches for a clear operation and clear the register.
6.  Perform the previous steps for other binary numbers.

*Fill-In Questions:*

1.  For data to be rotated correctly in a 4-bit register, _____ clock

    pulses are needed.

2.  Data that would normally be lost by a shift operation can be saved in the

    register by a simultaneous _____ operation.

3.  Data in a register can be inverted by a _____ rotate operation.

4.  To perform a complement rotate operation, the control input to the

    exclusive-OR gate must be a _____.

**SECTION 6-5
INSTANT REVIEW**

- A binary register is made up of flip-flops. The number of flip-flops in a register determine its bit word length. A 4-bit word is called a nibble and an 8-bit word is called a byte.
- The function of a register is to store data temporarily. A register can be designed to perform many operations with the data. These operations involve serial input and output, shifting right and left, parallel input and

output, clearing, rotating, complementing, serial-to-parallel conversion and parallel-to-serial conversion, and multiplication and division by factors of 2.

- Data may be transferred from one register to another by the serial mode or the parallel mode. The advantage of the serial mode is that fewer components are required, but its disadvantages are that clock pulses are needed, more time is required for shifting, and the data in the original register are usually lost. The advantages of parallel mode data transfer is that clock pulses are not needed, less time is required, and the data are not lost in the original register; however, its disadvantage is that more components are required. The compromise between these two transfer modes is called the trade-off factor.

## SECTION 6-6
## EXERCISES II

Perform all the exercises in this section before beginning the next section.

1.  List the function and operations a binary register is capable of performing.

    1.  Function:_____        2.  _____

    3.  _____        4.  _____

    5.  _____        6.  _____

    7.  _____        8.  _____

    9.  _____       10.  _____

    11.  _____       12.  _____

    13.  _____       14.  _____

Very often in digital circuits and computers, registers are arranged in 8-bit or byte form. The contents of these registers are displayed in the normal positional notation method with MBS to the far left and LSB to the far right, as shown in Figure 6-9. Most often, the hexadecimal number system is used to communicate with the digital system. Therefore, the registers are grouped by nibbles for easier usage. Reading right to left, the first 4 bits are a binary 11 or also hexadecimal B and the last 4 bits are a binary 4 or hexadecimal 4 (review Section 1-1h). The remaining exercises in this section will deal with 8-bit registers.

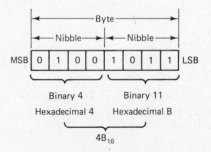

**Figure 6-9** Analyzing binary to hexadecimal conversion for one byte of data.

2. Grouping the contents of the registers into nibbles, indicate the hexadecimal number for the following 8-bit registers.

   a.  MSB 10100110 LSB = _____$_{16}$        b.  MSB 11111010 LSB = _____$_{16}$

   c.  MSB 10001100 LSB = _____$_{16}$        d.  MSB 00101110 LSB = _____$_{16}$

   e.  MSB 00111001 LSB = _____$_{16}$        f.  MSB 01111101 LSB = _____$_{16}$

3. Indicate the contents of the 8-bit registers with the hexadecimal numbers
given.

   a.  $24_{16}$ = MSB __ __ __ __ __ __ __ __ LSB

   b.  $A1_{16}$ = MSB __ __ __ __ __ __ __ __ LSB

   c.  $FF_{16}$ = MSB __ __ __ __ __ __ __ __ LSB

   d.  $69_{16}$ = MSB __ __ __ __ __ __ __ __ LSB

   e.  $CA_{16}$ = MSB __ __ __ __ __ __ __ __ LSB

   f.  $7A_{16}$ = MSB __ __ __ __ __ __ __ __ LSB

4. With the 8-bit register data given, indicate the hexadecimal number, then
complement the data and indicate the hexadecimal equivalent number.

   a.  MSB 01010111 LSB = _____$_{16}$

       complement = MSB __ __ __ __ __ __ __ __ LSB = _____$_{16}$

   b.  MSB 10101011 LSB = _____$_{16}$

       complement = MSB __ __ __ __ __ __ __ __ LSB = _____$_{16}$

   c.  MSB 01001111 LSB = _____$_{16}$

       complement = MSB __ __ __ __ __ __ __ __ LSB = _____$_{16}$

5. With the data given in the 8-bit register, indicate the hexadecimal number, indicate what the register contains after two shifts left, and indicate
the new hexadecimal equivalent number.

   a.  MSB 00111010 LSB = _____$_{16}$

       after 2 SHL = MSB __ __ __ __ __ __ __ __ LSB = _____$_{16}$

   b.  MSB 00001111 LSB = _____$_{16}$

       after 2 SHL = MSB __ __ __ __ __ __ __ __ LSB = _____$_{16}$

6. With the data given in the 8-bit register, indicate the hexadecimal num-

ber, indicate the contents of the register after two shifts right, and indicate the new hexadecimal equivalent number.

a.  MSB  11100110  LSB = _____ $_{16}$

after 2 SHR = MSB __ __ __ __ __ __ __ __  LSB = _____ $_{16}$

b.  MSB  10011010  LSB = _____ $_{16}$

after 2 SHR = MSB __ __ __ __ __ __ __ __  LSB = _____ $_{16}$

## SECTION 6-7
## TROUBLESHOOTING APPLICATION: TESTING THE BINARY REGISTER

If a register is going to be used in a digital system, all its features should be operating correctly. These features vary with different registers. Since the 7495 shift right/shift left register IC was used in three experiments of this unit, the testing procedures for this IC will be given here.

Connect the 7495 register IC to the logic switches and LEDs as shown in Figure 6-10a. If not enough switches are available, the data inputs $A$, $B$, $C$, and $D$ can be connected to a +5 V for a 1 and to ground for a 0 where indicated in the testing chart of Figure 6-10b.

Following the testing chart of Figure 6-10b, set the switches as indicated and record the output data appropriately under columns $Q_A$, $Q_B$, $Q_C$, and $Q_D$. The clock operation is indicated as ↑↓, which means go from 0 to 1 and then back to 0. The register will operate on the rising edge of the clock pulse.

## SECTION 6-8
## SELF-CHECKING QUIZZES

### 6-8a  BINARY REGISTER: TRUE-FALSE QUIZ

Place a T for true or an F for false to the left of each statement.

_____  1.  The function of a register is to provide temporary data storage.

_____  2.  A nibble is a group of 8 bits.

_____  3.  Registers are used to shift data left or right.

_____  4.  Parallel-to-serial conversion involves entering data in parallel and then shifting.

_____  5.  The Clr line to a register is used to reset it to zero.

_____  6.  A SHR operation shifts data to the preceding FFs.

_____  7.  Rotating data uses a serial output-to-serial input type of operation.

_____  8.  Parallel mode data transfer between registers uses fewer components than serial mode data transfer.

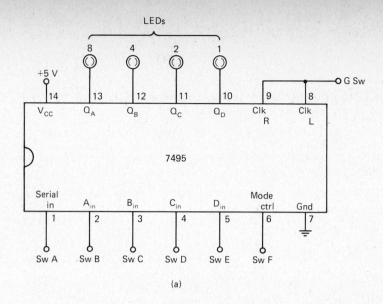

Figure 6-10 Testing the 7495 shift right/shift left register IC: (a) logic diagram; (b) testing chart.

| Operation | Sw A serial in | Sw B A$_{in}$ | Sw C B$_{in}$ | Sw D C$_{in}$ | Sw E D$_{in}$ | Sw F mode ctrl | Sw G Clk | Q$_A$ | Q$_B$ | Q$_C$ | Q$_D$ | Comments |
|---|---|---|---|---|---|---|---|---|---|---|---|---|
| Serial input | 1 | 0 | 0 | 0 | 0 | 0 | ↑↓ | | | | | FF$_A$ turns on |
| | 1 | 0 | 0 | 0 | 0 | 0 | ↑↓ | | | | | FF$_B$ turns on |
| | 1 | 0 | 0 | 0 | 0 | 0 | ↑↓ | | | | | FF$_C$ turns on |
| | 1 | 0 | 0 | 0 | 0 | 0 | ↑↓ | | | | | All FF$_s$ on |
| Shift right | 0 | 0 | 0 | 0 | 0 | 0 | ↑↓ | | | | | FF$_A$ turns off |
| | 0 | 0 | 0 | 0 | 0 | 0 | ↑↓ | | | | | FF$_B$ turns off |
| | 0 | 0 | 0 | 0 | 0 | 0 | ↑↓ | | | | | FF$_C$ turns off |
| | 0 | 0 | 0 | 0 | 0 | 0 | ↑↓ | | | | | All FF$_s$ off |
| At this point, connect the following: A$_{in}$-to-Q$_B$, B$_{in}$-to-Q$_C$, C$_{in}$-to-Q$_D$, D$_{in}$-to-Gnd. | | | | | | | | | | | | |
| Parallel inputs | 0 | 1 | 1 | 1 | 1 | 1 | ↑↓ | | | | | All FF$_s$ turns on |
| Shift Left | 0 | 0 | 0 | 0 | 0 | 1 | ↑↓ | | | | | FF$_D$ turns off |
| | 0 | 0 | 0 | 0 | 0 | 1 | ↑↓ | | | | | FF$_C$ turns off |
| | 0 | 0 | 0 | 0 | 0 | 1 | ↑↓ | | | | | FF$_B$ turns off |
| | 0 | 0 | 0 | 0 | 0 | 1 | ↑↓ | | | | | All FF$_s$ off |

(b)

_____ 9. Clock pulses are needed for a serial mode data transfer between registers.

_____ 10. A SHL operation shifts data to the following FFs.

## 6-8b  BINARY REGISTER: MULTIPLE-CHOICE QUIZ

Circle the correct answer for each question.

1. The number of clock pulses needed to serial load an 8-bit word register is:

   a. 1                     b. 2

   c. 4                     d. 8

2. A register contains the binary number MSB 11001100 LSB. After three shifts right the register contains:

    a. MSB 00110011 LSB      b. MSB 00011001 LSB

    c. MSB 01101100 LSB      d. MSB 00001100 LSB

3. A register contains the binary number MSB 10110101 LSB. After five shifts left the register contains:

    a. MSB 11010100 LSB      b. MSB 01010000 LSB

    c. MSB 10100000 LSB      d. MSB 01101010 LSB

4. The number of FFs used in a register to hold the binary equivalent of $153_{10}$ is:

    a. 4      b. 8

    c. 16      d. 32

5. A register contains the binary number MSB 10011101 LSB. After three shifts in a normal rotate operation, the register contains:

    a. MSB 01100111 LSB      b. MSB 11101000 LSB

    c. MSB 10110011 LSB      d. MSB 11010000 LSB

6. The mode of data transfer between two registers that uses less time is the:

    a. serial mode      b. parallel mode

7. A register contains the binary number MSB 11010111 LSB. After two shifts in a complement rotate operation, the register contains:

    a. MSB 00110101 LSB      b. MSB 00101000 LSB

    c. MSB 01101011 LSB      d. MSB 01000110 LSB

8. In a basic shift right register the type of flip-flop that *cannot* be used is:

    a. *J-K*      b. D-type

    c. T-type      d. answers a and b

    e. answers b and c

9. The number $5E_{16}$ would appear in a binary register as:

    a. MSB 11100101 LSB      b. MSB 01010111 LSB

    c. MSB 10100111 LSB      d. MSB 01011110 LSB

10. The inputs to the FFs that are connected together in some registers are:

    a. Clr      b. *K*

    c. Clk      d. Ps

    e. answers a and c      f. answers a and d

**ANSWERS TO EXPERIMENTS AND QUIZZES FOR UNIT 6**

*Experiment 1.*

(1) shifted     (2) $Q_A$, $Q_B$, $Q_C$, and $Q_D$     (3) $J$, $K$     (4) clock     (5) resets
(6) Ps     (7) serial-to-parallel, parallel-to-serial

*Experiment 2.*

(1) 1     (2) 0     (3) multiplication     (4) 1     (5) division     (6) 4
(7) 4

*Experiment 3.*

(1) logic     (2) 0100     (3) overall, task

*Experiment 4.*

(1) 4     (2) rotate     (3) complement     (4) 1

*True-False:*

(1) T     (2) F     (3) T     (4) T     (5) T     (6) F     (7) T     (8) F     (9) T
(10) F

*Multiple-Choice:*

(1) d     (2) b     (3) c     (4) b     (5) c     (6) b     (7) a     (8) c     (9) d
(10) e

# Unit 7

# Binary Counters

## SECTION 7-1
## IDENTIFICATION, THEORY, AND OPERATION

### 7-1a  INTRODUCTION

*Binary counters* consist of flip-flops connected together to produce output timing pulses which are used in other digital circuits. A basic counter will accept input pulses and give an indication of how many pulses have been received. Other counters will count down to produce a zero condition, provide a series of pulses from parallel outputs, and produce various timing sequences to operate other digital circuits.

There are numerous circuit configurations for various counters and to provide special counter features. This unit covers the operation of six basic counters:

1. Ring counter
2. Switch-tail ring counter
3. Binary up-counter
4. Binary down-counter
5. Synchronous counter
6. Other modulo counters

Mastering the techniques of these basic counters will enable you to understand other types of counter circuit configurations when encountered in digital equipment.

## 7-1b  RING COUNTER

A *ring counter* can be used to distribute a series of pulses entering a single input to parallel outputs in sequential order at various times. The ring counter can be produced from a shift right register using *J-K* flip-flops by connecting the outputs of the last flip-flop to the inputs of the first flip-flop as shown in Figure 7-1a. One of the flip-flops must be turned on (in this case $FF_A$ is preset) before the clock pulses begin; however, with added logic circuits some ring counters are self-starting.

With $FF_A$ on, a 1 appears at the *J* input of $FF_B$. The first clock pulse then turns on $FF_B$. Since the outputs of $FF_D$ are fed back to the inputs of $FF_A$, the *J* input is 0 and the *K* input is 1 because $Q_D$ is a 0 and $\overline{Q}_D$ is 1. Therefore, $FF_A$ turns off with the first clock pulse. With subsequent clock pulses the pulse is shifted along the counter, producing the various output pulses shown in the output timing chart of Figure 7-1b.

When $FF_D$ is on, its outputs are fed back to $FF_A$ to steer it on. The next clock pulse then turns on $FF_A$ as $FF_D$ is turned off and the cycle is repeated. With a relatively slow clock pulse rate, the pulse seems to "ripple" along the counter; hence these types of counters, where one flip-flip controls the following flip-flop in a serial manner, are called *ripple counters*. The ring counter can be used as a pulse distributor in a digital system by using the pulses at the *Q* outputs for other circuits. A low-going pulse on the clear line will reset or stop the counter action.

**Figure 7-1**  Ring counter: (a) logic diagram; (b) output timing chart.

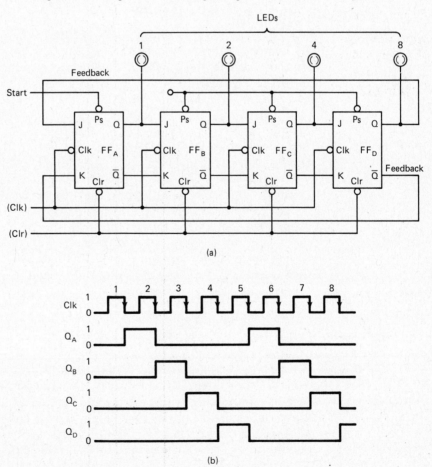

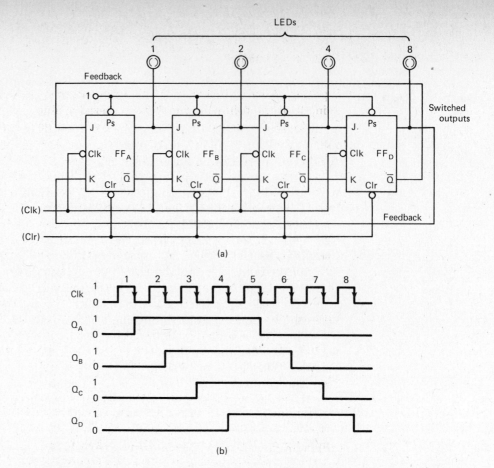

**Figure 7-2** Switch-tail ring counter: (a) logic diagram; (b) output timing chart.

### 7-1c  SWITCH-TAIL RING COUNTER

By switching the outputs of a ring counter the outputs of $FF_D$ are cross-coupled back to the inputs of $FF_A$ to produce a *switch-tail ring counter*, as shown in Figure 7-2a. This type of counter is also referred to as an *inverter ring counter* or *Johnson counter*.

The switch-tail ring counter is normally self-starting because when the counter is reset $Q_D$ is 0 and $\overline{Q}_D$ is 1, which steers $FF_A$ to turn on. When the first clock pulse appears, $FF_A$ turns on. With subsequent clock pulses each flip-flop turns on in succession. When all flip-flops are on, $Q_D$ is 1 and $\overline{Q}_D$ is 0, which steers $FF_A$ off. All the flip-flops are on for one input pulse cycle (from the trailing edge of input pulse 4 to the trailing edge of input pulse 5). As the clock pulses continue, each flip-flop then turns off in succession. The output waveforms for this counter are shown in Figure 7-2b. The switch-tail ring counter could be called a *sequential counter* since each flip-flop is turned on in succession and turned off in succession. A low-going pulse on the clear line will reset or stop the counter action, but if clock pulses are still present after the clear line goes high, the counter will begin its operation again.

### 7-1d  BINARY UP-COUNTER

A *binary up-counter* is used to tally or add up a series of input pulses, temporarily store them, perhaps display them, and use them in other operations. A basic 4-bit binary up-counter is constructed from T-type flip-flops as shown in Figure 7-3a. (Refer to Figures 5-9b and 5-10b for construction of a *J-K* flip-flop and D-type flip-flop, respectively.) The bubbles at the *T* inputs indicate that these flip-flops change states when their inputs go low. Notice that the $Q$ output of one flip-flop is connected to the $T$ input of the following flip-flop in a serial manner.

When the first pulse arriving at the input goes low, $FF_A$ turns on. Output $Q_A$ goes high, but does not affect $FF_B$. The counter contains a binary 1. When the second input pulse goes low, $FF_A$ turns off. Since $Q_A$ goes low, $FF_B$ turns on. Output $Q_B$ goes high, but does not affect $FF_C$. The counter now contains a binary 2. When the third input pulse goes low, $FF_A$ turns on, but $FF_B$ is not affected and remains on. Since $FF_B$ did not change states, no other flip-flops were affected and the counter contains a binary 3. When the fourth input pulse goes low, $FF_A$ turns off. Since output goes low, $FF_B$ turns off. Because output $Q_B$ goes low, $FF_C$ turns on and the counter contains a binary 4. This counting action continues until all flip-flops are on and the counter contains a binary 15. When the sixteenth input pulse goes low, $FF_A$ turns off. Since $Q_A$ goes low, $FF_B$ turns off. Similarly, $FF_C$ and $FF_D$ turn off and the counter is reset to zero, whereupon the counting cycle begins again if input pulses are present. The output timing pulses for this counting up operation are shown in Figure 7-3b. Of course, if the clear line goes low at any time, the counter will be reset to zero regardless of the number in the counter.

A binary up-counter can be used as a frequency divider. Notice that two complete input pulse cycles are needed to cause $FF_A$ to complete one output pulse cycle. Two complete output cycles of $FF_A$ will cause $FF_B$ to complete one cycle; however, it has taken four input pulses to achieve this. Therefore, each flip-flop in the counter divides the input frequency by a factor of 2. The output of each flip-flop is then $FF_A = f/2$, $FF_B = f/4$,

**Figure 7-3** Binary up-counter: (a) logic diagram; (b) output timing chart.

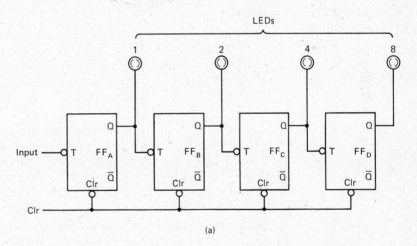

(a)

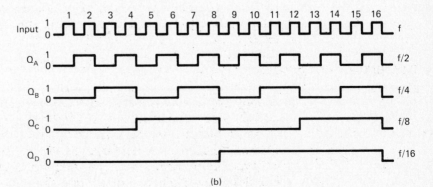

(b)

$FF_C = f/8$, and $FF_D = f/16$, where $f$ is the input pulse frequency. For example, if $f = 2000$ Hz, the output of the flip-flops is:

$$FF_A = 1000 \text{ Hz}$$

$$FF_B = 500 \text{ Hz}$$

$$FF_C = 250 \text{ Hz}$$

$$FF_D = 125 \text{ Hz}$$

This frequency division is indicated on the right-hand side of the output timing chart of Figure 7-3b.

The number of pulses that a counter may count depends on the number of flip-flops contained in the counter. This 4-bit binary up-counter will count up to 15 and reset to zero on the sixteenth input pulse. It is referred to as a *modulo 16 counter*. A 3-bit counter can count to 7 and then will reset on the eighth input pulse and is referred to as a *modulo 8 counter*. A 5-bit counter can count up to 31 and then will reset on the thirty-second input pulse and is referred to as a *modulo 32 counter*. Notice that the counter will count up or contain one less number than its assigned name. More information is given in Section 7-1g.

### 7-1e BINARY DOWN-COUNTER

With a basic *binary down-counter* the $\overline{Q}$ outputs are connected to the $T$ inputs of the flip-flops as shown in Figure 7-4a. The operation of the down-counter is the opposite of the up-counter. Initially, all flip-flops are turned

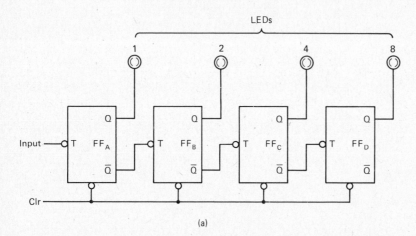

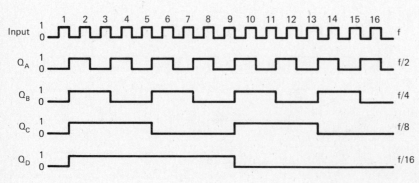

**Figure 7-4** Binary down-counter: (a) logic diagram; (b) output timing chart.

on and then one by one the contents of the counter is reduced to zero, where again the cycle is repeated if input pulses are present.

When the first input pulse goes low, $FF_A$ turns on. Output $\overline{Q}_A$ goes low and turns on $FF_B$. In a similar manner, $FF_C$ and $FF_D$ also turn on and the counter contains the binary 15. When the second input pulse goes low, $FF_A$ turns off, but since $\overline{Q}_A$ goes high no other flip-flops are affected and the counter contains a binary 14. When the third input pulse goes low, $FF_A$ turns on. Since $\overline{Q}_A$ goes low, $FF_B$ turns off, but because $\overline{Q}_B$ goes high, no other flip-flops are affected and the counter contains a binary 13. Continuing in a similar manner, the counter is reduced to zero by the sixteenth input pulse. On the seventeenth input pulse, all flip-flops turn on and the counting-down action begins again. These output timing pulses are shown in Figure 7-4b. Frequency division is also possible with this counter. Notice that the $\overline{Q}$ outputs of an up-counter resemble the down-counter operation and the $Q$ outputs of a down-counter resemble the up-counter operation.

## 7-1f  SYNCHRONOUS OPERATION

The counters in the previous sections are serial-type or ripple counters, where each flip-flop is controlled by a preceding flip-flop. In other words, a flip-flop cannot change states until the flip-flop before it changes states. At high operating frequencies, errors can occur in this type of counter because of the switching time required by each flip-flop. *Synchronous* (or *parallel*) *counters* can reduce or eliminate these errors in counting. A basic synchronous counter is shown in Figure 7-5a.

**Figure 7-5**  Synchronous counters: (a) basic synchronous counter; (b) 74193 4-bit synchronous up/down binary counter DIP IC.

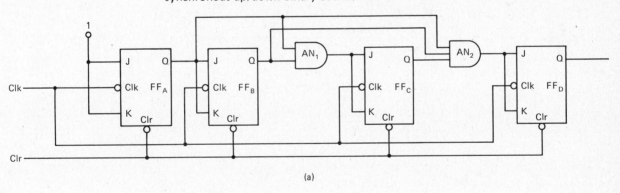

(a)

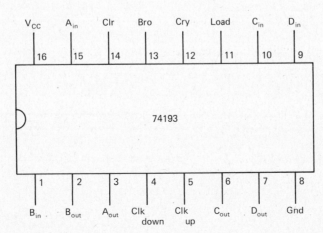

(b)

The essential requirement of a synchronous counter is that each flip-flop is driven by the system clock and any change in states occurs at the same time for all flip-flops. Logic gates are used to control the higher-order flip-flops, where the lower-order flip-flops must be on in order for the next flip-flop to turn on. For example, $FF_A$ and $FF_B$ must be on to cause $AN_1$ to produce a 1 at the $J$ input of $FF_C$. When the next input pulse (Clk) appears, $FF_C$ will turn on and not have to wait for $FF_A$ to turn off and then $FF_B$ to turn off. Similarly, $FF_A$, $FF_B$, and $FF_C$ have to be on to condition $AN_2$ with an output of 1 to turn on $FF_D$ with the eighth input pulse. As a result of this type of operation, the synchronous counter is faster than the ripple counter.

The pin identification is shown in Figure 7-5b for the versatile 74193 4-bit synchronous up/down binary counter IC. This counter has the capability of counting up or down, which is controlled by pins 4 and 5. Because of this feature, internal logic control gates are used to control the $T$ inputs of the flip-flops. The counter is also capable of being preset to any number from 0 through 15 by the parallel inputs at pins 1, 9, 10, and 15.

Pin 11 (load) is used to enable data to be entered at the parallel inputs. Parallel outputs are at pins 2, 3, 6, and 7. To count numbers greater than 15, subsequent ICs can be cascaded (connected in series). The carry (pin 12) of one IC is connected to the clock up (pin 5) of the following IC and the borrow (pin 13) is connected to the clock down (pin 4) to achieve a higher count. The input conditions for the various operations of this counter are as follows:

*To Preset the Counter*

1. Desired data are at inputs
2. Clear input = 0
3. Clock up input = 1
4. Clock down input = 1
5. Load input goes from high to low

*To Count Up*

1. Load input = 1
2. Clear input = 0
3. Clock down input = 1
4. Clock up input is pulsed

*To Count Down*

1. Load input = 1
2. Clear input = 0
3. Clock up input = 1
4. Clock down input is pulsed

*To Clear Counter*

1. Load input = 1
2. Clock up input = 1
3. Clock down input = 1
4. Clear input goes from low to high

## 7-1g OTHER MODULO COUNTERS

The modulus of a counter is the number of states through which it sequences during a cycle, or in other words, the magnitude of the count at which it resets. The 4-bit counters of the previous sections are classified as modulo 16 or simply MOD-16 counters. They counted to 15 and then reset on the sixteenth input pulse.

Counters can be wired to produce various count cycles. One of the easiest methods is to use a control gate connected to the counter reset line as shown in Figure 7-6a. The 7493 4-bit binary up-counter IC is one of several counter-type ICs available. A NAND gate, which is part of the IC, controls the resetting of the counter for various numbers. Notice that the output of $FF_A$ is not connected internally to the following flip-flop as the other flip-flops are. The first flip-flop alone then provides a divide-by-2, or MOD-2 counter. If the input pulses are applied to $FF_B$ at pin 1, the operation is a MOD-8 counter. When the output of $FF_A$ (pin 12) is connected to the input of $FF_B$ (pin 1), the operation becomes a MOD-16 counter. By connecting combinations of the outputs back to the inputs of the NAND gate, the counter can be used to count less than 16. Since the outputs of $FF_B$ and $FF_D$ are connected to the NAND gate, it becomes a MOD-10 counter, often called a *decade* or *BCD counter*. As the counting is proceeding, when $FF_B$ and $FF_D$ turn on, indicating a 10, both inputs to the NAND gate are 1, which produces a 0 at its output and resets the counter. As long as pulses

**Figure 7-6** Modulo counters: (a) typical 7493 4-bit binary up-counter wired as a MOD-10 counter; (b) divide-by-20 counter; (c) divide-by-60 counter.

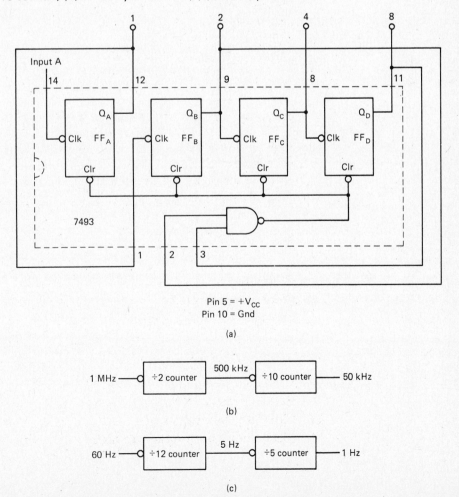

are present at the input, the counter will count to 9 and then reset on the tenth input pulse.

The 7493 IC can be wired for modulus counters from 2 to 15. However, when more than two outputs are fed back to the NAND gate, an external logic gate, such as an AND gate, is added to one of the NAND gate inputs to accept all the inputs.

There are other often used IC counters available, two of which are:

1.  7490 decade counter, which provides divide-by-2, 5, and 10 operation
2.  7492 divide-by-12 counter, which provides divide-by-2, 3, 6, and 12 operation

To determine how many flip-flops are required to count a particular number, simply convert the number to binary and count the number of bits. For example, to count the number $65_{10}$, convert it to binary:

$$
\begin{array}{lc}
 & \textit{Remainders} \\
65 \div 2 = 32 & 1 \\
32 \div 2 = 16 & 0 \\
16 \div 2 = 8 & 0 \\
8 \div 2 = 4 & 0 \quad\quad 1\ 0\ 0\ 0\ 0\ 0\ 1_2 \\
4 \div 2 = 2 & 0 \\
2 \div 2 = 1 & 0 \\
1 \div 2 = 0 & 1
\end{array}
$$

The binary number contains 7 bits; therefore, seven flip-flops are required. If the counter is allowed to count all bits, it is a MOD-128 counter, but with the reset control method it can be connected as a MOD-65 counter.

Low-number modulus counters can be cascaded together (connected in series) to produce larger-number frequency divisions, as shown in Figure 7-6b and c. A 1-MHz input frequency can be reduced by a divide-by-2 counter to 500 kHz and then a divide-by-10 counter to 50 kHz. The total division is then 20. Similarly, a 60-Hz input frequency can be reduced to 5 Hz by a divide-by-12 counter and then a divide-by-5 counter to 1 Hz, resulting in a total division of 60.

**SECTION 7-2
EXERCISES I**

Perform all the exercises in this section before beginning the next section.

1.  Draw the logic diagram for a 4-bit ring counter.

2.  Draw the output pulse timing chart for a 4-bit ring counter.

3. Draw the logic diagram for a 4-bit switch-tail ring counter.

4. Draw the output pulse timing chart for a 4-bit switch-tail ring counter.

5. Draw the logic diagram for a 4-bit binary up-counter.

6. Draw the output pulse timing chart for a 4-bit binary up-counter.

7. Draw the logic diagram for a 4-bit binary down-counter.

8. Draw the output pulse timing chart for a 4-bit binary down-counter.

9. Place an R for ripple counter or an S for synchronous counter to the left of each statement that describes that particular counter.

_____ a. This counter is the slowest in operation.

_____ b. The flip-flops in this counter change states at the same time.

_____ c. The $Q$ outputs are directly connected to the $T$ inputs.

_____ d. This counter is the fastest in operation.

_____ e. Preceding flip-flops must be on to condition logic gates to turn on the following flip-flop.

_____ f.  The counter with the fewest components.

_____ g.  The flip-flops in this counter change states in sequential order.

_____ h.  The counter with the most components.

_____ i.  This counter helps to reduce or eliminate counting errors.

10.  List the highest number that can be counted for the following counters.

   a.  MOD-8: _____     b.  MOD-16: _____     c.  MOD-32: _____

**SECTION 7-3**
**DEFINITION EXERCISES**

Give a brief description of each of the following terms.

1.  Ring counter _____

   _____

   _____

2.  Switch-tail ring counter _____

   _____

   _____

3.  Johnson counter _____

   _____

   _____

4.  Binary up-counter _____

   _____

   _____

5.  Binary down-counter _____

   _____

   _____

6.  MOD-10 counter _____

   _____

   _____

7.  Decade counter _____

_____

_____

8.  BCD counter _____

_____

_____

9.  Ripple counters _____

_____

_____

10.  Asynchronous counters _____

_____

_____

11.  Synchronous counters _____

_____

_____

12.  Frequency divider _____

_____

_____

13.  Modulus _____

_____

_____

## SECTION 7-4
## EXPERIMENTS

### EXPERIMENT 1.   RING COUNTER/SWITCH-TAIL RING COUNTER

*Objective:*

To demonstrate the operation of the ring counter and the switch-tail ring counter and observe the output timing pulses.

*Introduction:*

In this experiment you will produce an output timing pulse table for the ring counter and the switch-tail ring counter. An IC wiring diagram is given;

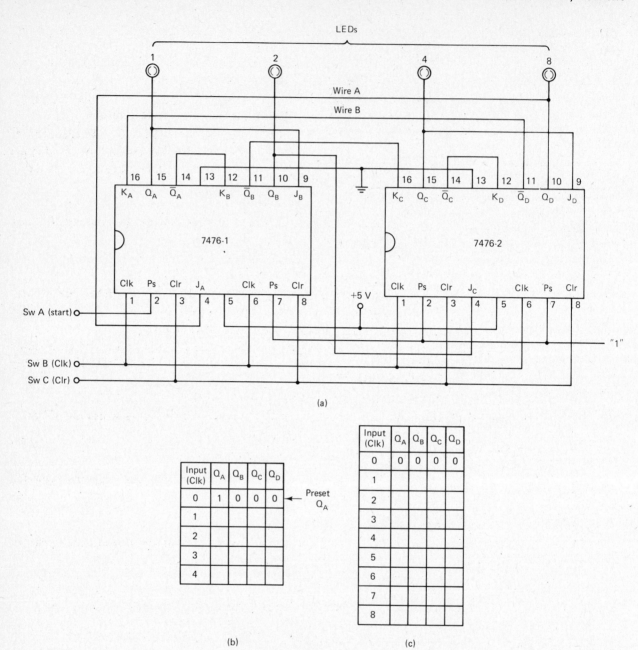

**Figure 7-7** Ring counter/switch-tail ring counter: (a) logic diagram; (b) ring counter output timing table; (c) switch-tail ring counter output timing table.

therefore, to better understand the logical operation of the counters, refer to Figures 7-1 and 7-2. The circuit shown in Figure 7-7a is used for both types of counters, and only two wires are changed to convert from one to the other.

*Materials Needed:*

1   Digital logic trainer *or*

2   7476 dual *J-K* master/slave flip-flop DIP ICs

4   LED indicators

    Several hookup wires or leads

*Procedure:*

### Ring Counter

1. Construct the circuit shown in Figure 7-7a.
2. Place Sw $A$, Sw $B$, and Sw $C$ high.
3. Move Sw $C$ (Clr) low and then high to reset the counter.
4. Move Sw $A$ (start) low and then high to preset $FF_A$. $Q_A$ should turn on.
5. Move Sw $B$ (Clk) low and then high and record the output conditions of the counter in the appropriate place in the output timing table shown in Figure 7-7b.
6. Repeat step 5 three more times.
7. You can replace Sw $B$ with a slow-running clock to get the visual effect of a ripple counter.

### Switch-Tail Ring Counter

8. Exchange or switch wire $A$ from pin 10 to pin 11 on IC 7476-2 and wire pin $B$ from pin 11 to pin 10 on IC 7476-2.
9. Move Sw $C$ (Clr) low and then high to reset counter.
10. Move Sw $B$ (Clk) low and then high and record the output conditions of the counter in the appropriate place in the output timing table shown in Figure 7-7c.
11. Repeat step 10 seven more times.
12. Replacing Sw $B$ with a slow-running clock produces a unique effect of a sequential ripple counter.

*Fill-In Questions:*

1. It takes _____ clock pulses to complete one counting cycle for a 4-bit ring counter.

2. It takes _____ clock pulses to complete one counting cycle for a 4-bit switch-tail ring counter.

3. The ring counter can be used as a pulse _____.

4. Both of these counters are _____ counters.

## EXPERIMENT 2.  BINARY UP/DOWN-COUNTERS

*Objective:*

To show the counting effects of an up-counter and a down-counter and show how counters are used as frequency dividers.

*Introduction:*

This experiment uses *J-K* flip-flops wired as T-type flip-flops in the counters. Only the IC connections are shown and you should refer to Figures 7-3 and 7-4 for the logic diagram analysis. The circuit shown in Figure 7-8a is used for both types of counters, and only three wires are changed to convert from one to the other. If it is preferred, you can use the 74193 synchronous up/down-counter IC shown in Figure 7-5b and apply the operations given in Section 7-1f.

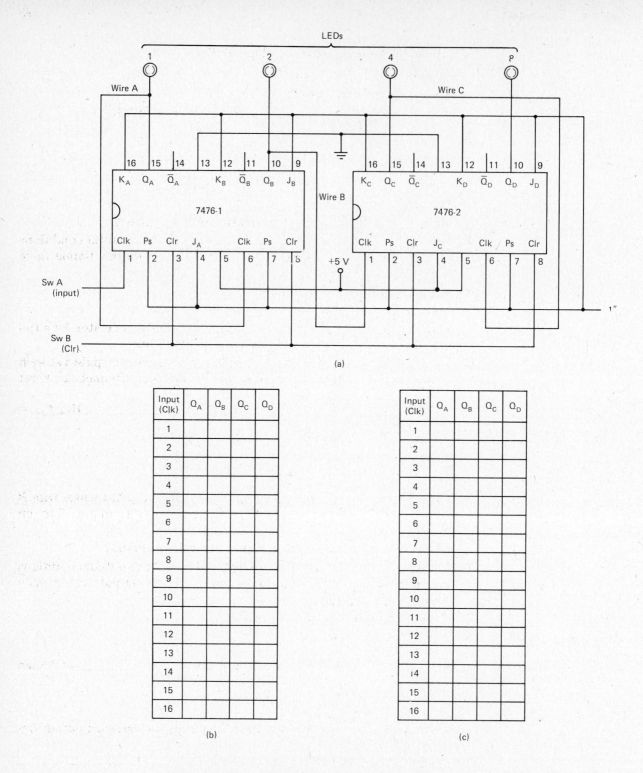

**Figure 7-8** Binary up/down-counters: (a) logic diagram; (b) up-counter output timing table; (c) down-counter output timing table.

*Materials Needed:*

*1* Digital logic trainer *or*

*1* Square-wave pulse generator

*1* Dual-trace oscilloscope

   2  7476 dual *J-K* master/slave flip-flop DIP ICs

   4  LED indicators

     Several hookup wires or leads

     Optional: *1*  74193 synchronous up/down-counter DIP IC

*Procedure:*

### Up-Counter

1. Construct the circuit shown in Figure 7-8a.
2. Place Sw *A* and Sw *B* high.
3. Move Sw *B* (Clr) low and then high to reset the counter.
4. Move Sw *A* (input) low and then high and record the output conditions of the counter in the appropriate place in the output timing table shown in Figure 7-8b.
5. Repeat step 4 fifteen more times.
6. Replace Sw *A* with the square-wave pulse generator.
7. Using one channel of the oscilloscope, set the pulse generator for a frequency of 10 kHz with a 3-V peak-to-peak amplitude.
8. Using the second channel of the oscilloscope, observe the pulses at each output and calculate the frequency. Record these frequencies here:

$f_{in}$ = 10 kHz, $f_{Q_A}$ = _____ Hz, $f_{Q_B}$ = _____ Hz, $f_{Q_C}$ =

_____ Hz, $f_{Q_D}$ = _____ Hz.

### Down-Counter

9. Change wire *A* from pin 15 to pin 14 on IC 7476-1. Change wire *B* from pin 10 to pin 11 on IC 6476-1. Change wire *C* from pin 15 to pin 14 on IC 7476-2.
10. Move Sw *B* (Clr) low and then high to reset the counter.
11. Move Sw *A* (input) low and then high and record the output conditions of the counter in the appropriate place in the output timing table shown in Figure 7-8c.
12. Repeat step 8 fifteen more times.

*Fill-In Questions:*

1. A 4-bit binary up-counter will count up to _____ and then

   reset on the _____ input pulse.

2. When a 4-bit up-counter is wired as a frequency divider, each output frequency can be expressed as _____, _____, _____

   _____, and _____.

3. All the flip-flops in a down-counter will turn on with the _____

   input pulse and then count down to _____.

4. These two ripple counters use _____-type flip-flops.

**EXPERIMENT 3.  MODULO COUNTERS**

*Objective:*

To demonstrate how to construct a MOD-10 and MOD-13 counter from a 4-bit binary counter.

*Introduction:*

This experiment produces a MOD-10 counter using the internal NAND gate of the 7493 IC and also shows how to wire a MOD-13 counter using an external AND gate connected to one input of the internal NAND gate.

*Materials Needed:*

1   Digital logic trainer *or*
1   7493 4-bit binary counter DIP IC
1   7408 quad two-input AND gate DIP IC
4   LED indicators
    Several hookup wires or leads

*Procedure:*

*MOD-10 Counter*

1.  Construct the circuit shown in Figure 7-9a. Make sure that all LEDs are initially off.
2.  Activate Sw *A* while keeping track of how many pulses are entered into the counter and monitor the LEDs for the numerical count.

3.  How many pulses were entered to reset the counter? _____

4.  What was the highest number in the counter before it reset to zero?

    _____ Which LEDs were on to indicate this number? _____

    _____ and _____.

*MOD-13 Counter*

5.  Construct the circuit shown in Figure 7-9b. Make sure that all LEDs are initially off.
6.  Activate Sw *A* while keeping track of how many pulses are entered into the counter and monitor the LEDs for the numerical count.

7.  How many pulses were entered to reset the counter? _____

8.  What was the highest number in the counter before it reset to zero?

    _____ Which LEDs were on to indicate this number? _____

    _____ and _____.

*Fill-In Questions:*

1.  A MOD-10 counter will reset after _____ input pulses.

2.  The highest number contained in a MOD-10 counter before reset is

    _____.

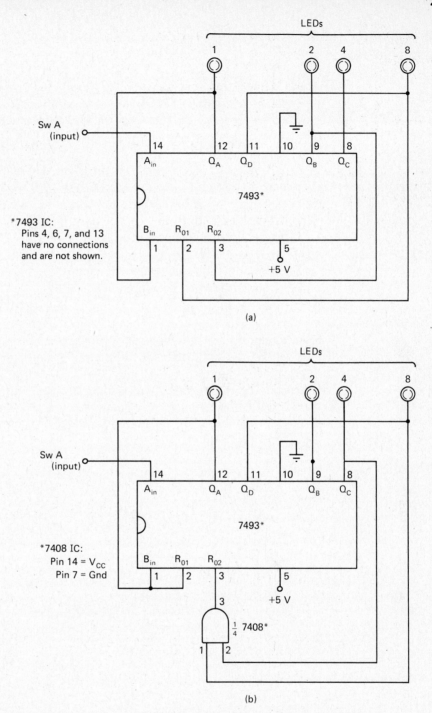

**Figure 7-9** Modulo counters: (a) MOD-10 counter logic diagram; (b) MOD-13 counter logic diagram.

3. A MOD-13 counter will reset after _____ input pulses.

4. The highest number contained in a MOD-13 counter before reset is

   _____.

5. These two types of counters used the _____ method of control.

**SECTION 7-5**
**INSTANT REVIEW**

- A ring counter is basically a shift right register with its outputs connected back to its inputs. It can be used as a pulse distributor where a series of pulses entering its input are applied to parallel outputs at different times.
- The switch-tail ring counter is the same as a ring counter, except that the outputs are cross-connected or switched back to the inputs. Its operation is sequential; each flip-flop turns on, one after another, until all are on and then each flip-flop turns off, one after another, until all are off.
- A binary up-counter adds up or tallies the pulses applied to its input. When all flip-flops are on, the next input pulse resets the counter to zero. The count cycle then begins again.
- With a binary down-counter, all flip-flops turn on with the first input pulse and subsequent pulses reduce the contents of the counter to zero. The count-down cycle then begins again.
- The modulus of a counter is the highest number of input pulses that causes it to reset.
- Counters are also used as frequency dividers. The output frequency of each flip-flop is one-half of its input frequency.
- Asynchronous counters, or ripple counters, have flip-flops that change state by the action of a preceding flip-flop, whereas synchronous counters have additional gating circuits and all flip-flop action is accomplished simultaneously by clock pulses. Ripple counters are slower than synchronous counters.

**SECTION 7-6**
**EXERCISES II**

Perform all the exercises in this section before beginning the next section.

1. If a ring counter has 4 bits and only $Q_A$ is on, then after six shifts, indicate the condition of the following outputs:

   $Q_A =$_____, $Q_B =$_____, $Q_C =$_____, $Q_D =$_____

2. If a switch-tail ring counter has 4 bits and only $Q_A$ is on, then after seven shifts, indicate the conditions of the following outputs:

   $Q_A =$_____, $Q_B =$_____, $Q_C =$_____, $Q_D =$_____

3. A 4-bit up-counter has a pulse train of 18 pulses applied to its input. After the pulse train ends, the conditions of the outputs are:

   $Q_A =$_____, $Q_B =$_____, $Q_C =$_____, $Q_D =$_____

4. A 4-bit down-counter has a pulse train of 20 pulses applied to its input. After the pulse train ends, the conditions of the outputs are:

   $Q_A =$_____, $Q_B =$_____, $Q_C =$_____, $Q_D =$_____

5. A MOD-12 counter with reset has a pulse train of 15 pulses applied to its input. After the pulse train ends, the conditions at the outputs are:

$Q_A =$ _____, $Q_B =$ _____, $Q_C =$ _____, $Q_D =$ _____

6. List the number of flip-flops needed in a counter to divide the following frequencies:

    a.  $f_{in} = 160$ kHz            b.  $f_{in} = 256$ kHz
        $f_{out} = 5$ kHz               $f_{out} = 1$ kHz

      FFs = _____         FFs = _____

7. List the output frequencies for the following counters:

    a.  6-bit counter            b.  8-bit counter
        $f_{in} = 4$ MHz              $f_{in} = 32$ kHz

      $f_{out} =$ _____     $f_{out} =$ _____

8. List the number of flip-flops needed for the following modulo counters:

    a.  MOD-24 counter         b.  MOD-100 counter

      FFs = _____         FFs = _____

9. Using modulo counters from 2 to 16, draw the cascaded counters needed to reduce the input frequencies to the output frequencies given below. List the divided frequency after each counter and the total division required. Refer to Figure 7-6a and b as a guide. (*Hint:* Divide the $f_{in}$ by the $f_{out}$ to find the total division.)

| $f_{in}$ | $f_{out}$ | *Total Division* |
|---|---|---|
| a.  2 MHz | 100 kHz | |
| b.  150 kHz | 5 kHz | |
| c.  36 kHz | 2 kHz | |
| d.  120 Hz | 1 Hz | |

10. Draw a MOD-11 counter with reset logic diagram using the 7493 IC shown in Figure 7-6a. (*Hint:* An additional external logic gate may be needed.)

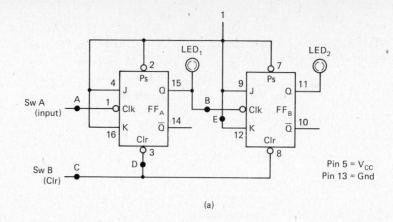

(a)

| Condition | Input pulse | $Q_A$ | $Q_B$ | Comments |
|---|---|---|---|---|
| Normal | 1 | | | 1st input pulse turns on FF |
| | 2 | | | 2nd input pulse turns off $FF_A$ and $FF_B$ turns on |
| | 3 | | | 3rd input pulse turns on $FF_A$ and $FF_B$ remains on |
| | 4 | | | 4th input pulse turns off $FF_A$ and $FF_B$ turns off |
| Point A open | 1-3 | | | Input pulses not reaching counter |
| Point B open | 1 | | | 1st input pulse turns on $FF_A$ |
| | 2 | | | 2nd input pulse: $FF_A$ turns off, $FF_B$ remains off |
| | 3 | | | 3rd input pulse turns on $FF_A$, $FF_B$ remains off |
| | 4 | | | 4th input pulse: $FF_A$ turns off, $FF_B$ remains off |
| Point C open | 1-3 | | | Counter fully loaded |
| Clear counter with Sw B | | | | Counter remains the same, will not reset |
| Point D open | 1-3 | | | Counter fully loaded |
| Clear counter with Sw B | | | | $FF_A$ remains on, $FF_B$ resets |
| Open point E and short input K to ground | 1-3 | | | Count up okay |
| | 4 | | | $FF_A$ turns off, but $FF_B$ remains on |

(b)

**Figure 7-10**  Troubleshooting an up-counter: (a) logic diagram; (b) trouble table.

## SECTION 7-7
## TROUBLESHOOTING APPLICATION: TESTING A 2-BIT BINARY UP-COUNTER

Unless the IC is malfunctioning itself, many other problems with counters can be caused by the input lines to it. When ICs are mounted into sockets, one or two pins could become bent and not make an electrical contact with the circuit. This basic counter can provide the necessary troubleshooting skills to test other type counters. Construct the circuit shown in Figure 7-10a. A logic pulser can be used to apply pulses to the input and a logic probe can be used to check the Q outputs.

Using the trouble table in Figure 7-10b, set the trouble condition, enter the pulses indicated in the input pulse column and then record the conditions of the outputs in columns $Q_A$ and $Q_B$. When the input pulse column reads 1-3, enter three pulses before recording the outputs. The next input pulse or operation will indicate the problem.

**SECTION 7-8**
**SELF-CHECKING QUIZZES**

**7-8a  BINARY COUNTERS: TRUE-FALSE QUIZ**

Place a T for true or an F for false to the left of each statement.

_____ 1. A MOD-12 counter contains 12 flip-flops.

_____ 2. A binary up-counter turns on all the flip-flops with the first input pulse.

_____ 3. A MOD-4 counter will divide the input frequency by 4.

_____ 4. A ring counter can be used for pulse distribution.

_____ 5. A synchronous counter requires more components and is slower in operation.

_____ 6. *J-K*, T-type, and D-type flip-flops can be used in counters.

_____ 7. In a switch-tail ring counter, all flip-flops are on for at least one clock pulse cycle.

_____ 8. A 4-bit binary up-counter requires 16 input pulses to complete one count cycle.

_____ 9. A divide-by-3 counter is cascaded with a divide-by-6 counter. With an input frequency of 81 kHz, the output frequency would be 9 kHz.

_____ 10. To count the number $255_{10}$, a counter needs eight flip-flops.

**7-8b  BINARY COUNTERS: MULTIPLE-CHOICE QUIZ**

Circle the correct answer for each question.

1. A 6-bit binary up-counter would reset on input pulse:

   a. 6        b. 16
   c. 32      d. 64

2. A 4-bit up-counter is normally reset. After 21 input pulses, the conditions of the outputs are:

   a. $Q_A = 0$      b. $Q_A = 1$
      $Q_B = 0$           $Q_B = 0$
      $Q_C = 0$           $Q_C = 1$
      $Q_D = 0$           $Q_D = 0$

   c. $Q_A = 0$      d. $Q_A = 0$
      $Q_B = 1$           $Q_B = 1$
      $Q_C = 0$           $Q_C = 1$
      $Q_D = 1$           $Q_D = 0$

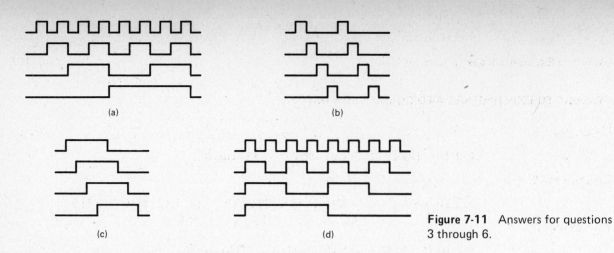

(a)

(b)

(c)

(d)

**Figure 7-11** Answers for questions 3 through 6.

For questions 3 through 6, match the type of counter with the output timing charts used for answers shown in Figure 7-11.

3. Ring counter

   (a)    (b)    (c)    (d)

4. Switch-tail ring counter

   (a)    (b)    (c)    (d)

5. Binary up-counter

   (a)    (b)    (c)    (d)

6. Binary down-counter

   (a)    (b)    (c)    (d)

7. Synchronous counters:

   a. have faster operating speed

   b. reduce counting errors

   c. both answers a and b

   d. none of the above

8. A 6-bit ripple up-counter with an input frequency of 100 kHz would have an output frequency on the fourth flip-flop of:

   a. 25 kHz          b. 12.5 kHz

   c. 6.25 kHz        d. 3.125 kHz

9. Refer to Figure 7-12. Connecting the outputs of $Q_A$ and $Q_D$ to the inputs of the NAND gate (pins 2 and 3), respectively, produces a:

   a. MOD-8 counter      b. MOD-9 counter

   c. MOD-11 counter     d. MOD-12 counter

10. Refer to Figure 7-12. Connecting the outputs of $Q_C$ and $Q_D$ to the inputs of the NAND gate (pins 2 and 3), respectively, produces a:

    a. MOD-8 counter      b. MOD-9 counter

    c. MOD-11 counter     d. MOD-12 counter

**Figure 7-12** For questions 9 and 10.

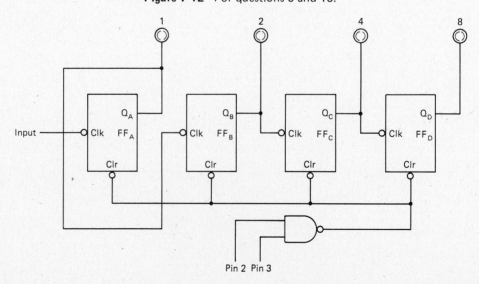

## ANSWERS TO EXPERIMENTS AND QUIZZES FOR UNIT 7

*Experiment 1.*

(1) 4     (2) 8     (3) distributor     (4) ripple

*Experiment 2.*

(1) 15, sixteenth     (2) $f/2, f/4, f/8, f/16$     (3) first, zero     (4) $T$

*Experiment 3.*

(1) 10     (2) 9     (3) 13     (4) 12     (5) reset

*True-False:*

(1) F     (2) F     (3) T     (4) T     (5) F     (6) T     (7) T     (8) T     (9) F
(10) T

*Multiple-Choice:*

(1) d     (2) b     (3) b     (4) c     (5) a     (6) d     (7) c     (8) c     (9) b
(10) d

# Unit 8

---

# Arithmetic Circuits

## SECTION 8-1
## IDENTIFICATION, THEORY, AND OPERATION

### 8-1a  INTRODUCTION

Digital electronic circuits involve numbers in the form of pulses; therefore, various logic circuits can be connected together to perform mathematical operations. The four basic mathematical operations are adding, subtracting, multiplying, and dividing. Separate logic circuits can be designed for each of these operations; however, with the use of complementation, shifting, and comparing data as discussed in the preceding units, only an adder circuit is required to perform all these operations and more.

Multiplication is successive addition, and division is essentially successive subtraction; therefore, this unit will deal with the principles of binary addition and subtraction and their associated circuits.

### 8-1b  BINARY ADDITION

Binary addition is similar to adding in decimal and there are four basic rules:

$$
\begin{array}{llll}
1. \quad \begin{array}{r} 0 \\ +\,0 \\ \hline 0 \end{array} &
2. \quad \begin{array}{r} 0 \\ +\,1 \\ \hline 1 \end{array} &
3. \quad \begin{array}{r} 1 \\ +\,0 \\ \hline 1 \end{array} &
4. \quad \text{Carry} \quad 1 \\
\end{array}
$$

4. Carry $1$
   $1 \quad +1$
   $\longrightarrow 10$  (with a carry of 1)

The first three rules are the same as in decimal, but rule 4 may appear confusing. This rule reads: "1 plus 1 equals 0 with a carry of 1." Remember that $2_{10} = 10_2$; therefore, $1_2 + 1_2 = 10_2$ (2 in binary).

208

When larger binary numbers are added, the resulting 1s are carried into the next-higher-order positions. The following examples show binary addition and demonstrate the carry principle. The binary equivalent is shown in normal positional notation with LSB to the extreme right, and addition is performed from right to left beginning with the first column (LSB). Notice the mathematical terminology of augend, addend, and sum.

**Example 1:**

*Decimal*                          *Binary equivalent*

```
                                   1 ◄─── Carry
  1 0  (augend)                   1 0 1 0
+ 0 3  (addend)                 + 0 0 1 1
  1 3  (sum)                      1 1 0 1
```

Beginning with the first column (LSB) in Example 1, "0 plus 1 equals 1"; in the second column, "1 plus 1 equals 0 with a carry of 1 to the next column"; in the third column, "the carry 1 plus 0 equals 1 plus 0 equals 1"; and in the fourth column, "1 plus 0 equals 1."

**Example 2:**

*Decimal*                          *Binary equivalent*

```
                                          Carry 2
  1 ◄─── Carry                     1 1 ◄─── Carry 1
  0 9                              1 0 0 1
                                       1 ◄─── Subtotal
+ 0 3                            + 0 0 1 1
  1 2                              1 1 0 0
```

Beginning with the first column in Example 2, "1 plus 1 equals 0 with a carry of 1 to the next column"; in the second column, "the carry 1 plus 0 equals 1." At this point a subtotal of 1 can be assumed, which is added to the remaining 1 in the second column; therefore, "1 plus 1 equals 0 with a carry of 1 to the third column." In the third column, "the carry of 1 plus 0 equals 1 plus 0 equals 1"; and in the fourth column, "1 plus 0 equals 1."

**Example 3:**

*Decimal*                          *Binary equivalent*

```
                                          Carry 2
                                   1 1 ◄─── Carry 1
  1 1                              1 0 1 1
                                       0 ◄─── Subtotal
+ 0 3                            + 0 0 1 1
  1 4                              1 1 1 0
```

In Example 3 there will be three 1s to be added together in the second column. Beginning with the first column, "1 plus 1 equals 0 with a carry of 1 to the next column"; in the second column, "1 plus 1 equals a subtotal of 0

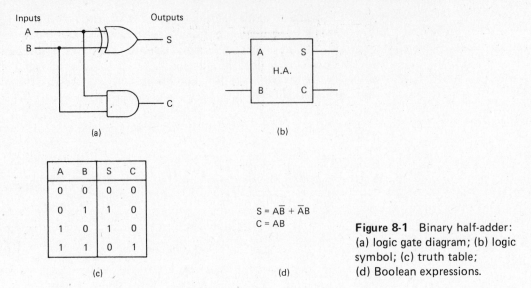

**Figure 8-1**  Binary half-adder:
(a) logic gate diagram; (b) logic
symbol; (c) truth table;
(d) Boolean expressions.

with a carry of 1 to the next column" (the subtotal 0 plus 1 equals 1 for the
second column); in the third column, "the carry 1 plus 0 equals 1 plus 0
equals 1"; and in the fourth column, "1 plus 0 equals 1."

### 8-1c  BINARY HALF-ADDER

Various logic gates can be connected together to produce a half-adder, but
one of the simplest is an exclusive-OR gate and AND gate as shown in Figure 8-1a. These logic gates can be represented by the logic symbol shown
in Figure 8-1b. This circuit can be used to add one column of numbers or
verify the four rules of binary addition. The augend is at input $A$ and the
addend is at input $B$. The outputs are $S$ for sum and $C$ for carry. The truth
table shown in Figure 8-1c verifies the four rules of binary addition. When
a 1 appears at any output, examining the input conditions provides a means
of writing a Boolean expression for each output as shown in Figure 8-1d.
The following summary will help in remembering the operation of a half-
adder.

1. When both inputs are 0: $S = 0$ and $C = 0$.
2. When only one input is 1: $S = 1$ and $C = 0$.
3. When both inputs are 1: $S = 0$ and $C = 1$.

### 8-1d  BINARY FULL-ADDER

Since a binary half-adder is limited to a single column of two numbers, a
binary full-adder, shown in Figure 8-2a, is required to handle multiple-
column numbers with a provision for a carry-in from the preceding columns.
A binary full-adder consists of two half-adders and an OR gate, as shown in
Figure 8-2b. The augend is at input $A$, the addend is at input $B$, and a carry-
in resulting from a previous column is at input $C_{\text{in}}$. The outputs are sum-out
$S_o$, and carry-out $C_o$. The binary full-adder logic symbol most often used is
shown in Figure 8-2c. Conditions at the outputs depend on the conditions
present at the inputs, as shown by the truth table in Figure 8-2d. Any time
an output is 1, the input conditions can be listed to give the Boolean expression, as shown in Figure 8-2e. The following summary will help to familiarize
you with the operations of a binary full-adder.

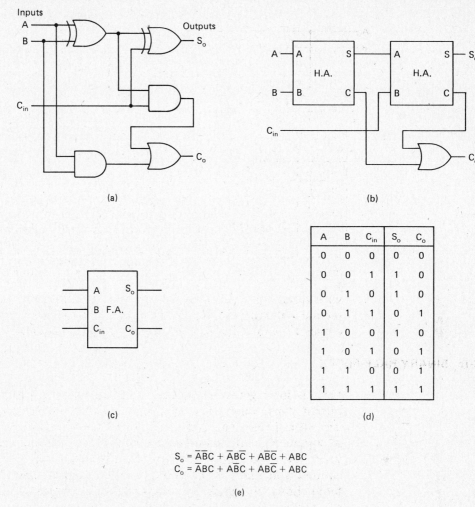

**Figure 8-2**  Binary full-adder: (a) logic gate diagram; (b) block diagram; (c) logic symbol; (d) truth table; (e) Boolean expressions.

1. When all inputs are 0, $S_o = 0$ and $C_o = 0$.
2. When any one input is 1, $S_o = 1$ and $C_o = 0$.
3. When any two inputs are 1, $S_o = 0$ and $C_o = 1$.
4. When all inputs are 1, $S_o = 1$ and $C_o = 1$.

## 8-1e  BINARY SUBTRACTION

Binary subtraction is similar to subtracting in decimal and there are four basic rules:

$$
\begin{array}{llll}
1. \quad \begin{array}{r} 0 \\ -\,0 \\ \hline 0 \end{array} &
2. \quad \begin{array}{r} 1 \\ -\,0 \\ \hline 1 \end{array} &
3. \quad \begin{array}{r} 1 \\ -\,1 \\ \hline 0 \end{array} &
4. \quad \begin{array}{r} 0 \\ -\,1 \\ \hline 1 \end{array} \text{ (with a borrow of 1)}
\end{array}
$$

The first three rules are the same as for decimal, but rule 4 is probably the most confusing. Obviously, you cannot subtract 1 from 0 unless there is a 1 in the next-higher-order columns to borrow from. The following examples will show binary subtraction and demonstrate the borrow principle. Notice the mathematical terminology of minuend, subtrahend, and difference.

**Example 1:**

*Decimal*                                            *Binary equivalent*

$$
\begin{array}{rll}
 & 1\ 1 & \text{(minuend)} \\
- & 0\ 6 & \text{(subtrahend)} \\
\hline
 & 0\ 5 & \text{(difference)}
\end{array}
$$

Beginning with the first column (LSB) in Example 1, "1 minus 0 equals 1"; in the second column, "1 minus 1 equals 0"; and in the third column, "0 minus 1 requires a borrow from the fourth column." When you borrow a 1 from the fourth column, a 0 is left in this column and the third column becomes (10), a binary 2 (the parentheses may help you remember that it is a binary 2); therefore, "10 minus 1 equals 1," and in the fourth column, "0 minus 0 equals 0."

**Example 2:**

*Decimal*                                            *Binary equivalent*

$$
\begin{array}{rl}
 & 9 \\
- & 3 \\
\hline
 & 6
\end{array}
$$

Beginning with the first column in Example 2, "1 minus 1 equals 0"; in the second column, "0 minus 1 requires a borrow from the third column, which does not contain a 1; therefore, you must go to the fourth column to borrow a 1, which leaves a 0." Column three is now a binary 2 (10). Borrowing 1 from the third column leaves a 1 and the second column now contains a binary 2 (10); therefore, "10 minus 1 equals 1"; in the third column, "1 minus 0 equals 1"; and finally in the fourth column, "0 minus 0 equals 0."

## 8-1f  BINARY SUBTRACTION USING COMPLEMENTATION

Binary subtracter circuits can be constructed from logic gates; however, another method of subtracting which uses less digital circuitry is by complementing and adding. This method is preferred and used in microprocessors where only an adder is required.

As you recall from the study of flip-flops, the complement of a binary number is simply its inversion or opposite state. For example, the *complement* of 0 is 1 and the complement of 1 is 0. In Section 6-1d.2 it was shown how a register could be complemented by using an exclusive-OR gate and rotating the data through it back to the input of the register (refer to Figure 6-3).

### 8-1f.1  The 1's-Complement Method of Subtraction

The *1's-complement method* of subtraction is to invert or complement the subtrahend, add it to the minuend, and then again add the last carry to the difference to get the correct answer. The following examples show this operation.

**Example 1:**

*Normal subtraction*                                                    *1's-complement method*

$$
\begin{array}{cccc}
 & 0 & (1) & \\
0 & \cancel{1} & (0) & 1 \\
-0 & 0 & 1 & 1 \\
\hline
0 & 0 & 1 & 0
\end{array}
\quad \Rightarrow 1\text{'s complement} =
\quad
\begin{array}{ccccc}
1 & 1 & & & \\
 & 0 & 1 & 0 & 1 \\
+ & 1 & 1 & 0 & 0 \\
\hline
1 & 0 & 0 & 0 & 1
\end{array}
$$

Add last carry $\longrightarrow +1$

$$
\begin{array}{cccc}
0 & 0 & 1 & 0
\end{array}
$$

**Example 2:**

*Normal subtraction*                                                    *1's-complement method*

$$
\begin{array}{cccc}
0 & (1) & & \\
\cancel{1} & (0) & 1 & 1 \\
-0 & 1 & 1 & 0 \\
\hline
0 & 1 & 0 & 1
\end{array}
\quad \Rightarrow 1\text{'s complement} =
\quad
\begin{array}{ccccc}
1 & & 1 & 1 & \\
1 & 0 & 1 & 1 & \\
+ & 1 & 0 & 0 & 1 \\
\hline
1 & 0 & 1 & 0 & 0
\end{array}
$$

Add last carry $\longrightarrow +1$

$$
\begin{array}{cccc}
0 & 1 & 0 & 1
\end{array}
$$

### 8-1f.2 The 2's-Complement Method of Subtraction

The *2's-complement method* of subtraction is actually just as easy as the 1's-complement method and is used in most microprocessor applications. First, find the 1's complement of the subtrahend and then add 1 to it to get the 2's complement. Now add the complemented subtrahend to the minuend, but drop or ignore the last carry to get the correct difference. The following examples show this operation.

**Example 1:**

*Normal subtraction*                                          *2's-complement method*

$$
\begin{array}{cccc}
 & 0 & (1) & \\
0 & \cancel{1} & (0) & 1 \\
-0 & 0 & 1 & 1 \\
\hline
0 & 0 & 1 & 0
\end{array}
\quad = 1\text{'s comp} = 1\ 1\ 0\ 0
$$

$\Rightarrow 2\text{'s comp } (+1) =$

$$
\begin{array}{ccccc}
1 & 1 & 1 & & \\
 & 0 & 1 & 0 & 1 \\
+ & 1 & 1 & 0 & 1 \\
\hline
1 & 0 & 0 & 1 & 0
\end{array}
$$

$\downarrow$

Drop last carry

**Example 2:**

*Normal subtraction*                                          *2'-complement method*

$$
\begin{array}{cccc}
0 & (1) & & \\
\cancel{1} & (0) & 1 & 1 \\
-0 & 1 & 1 & 0 \\
\hline
0 & 1 & 0 & 1
\end{array}
\quad = 1\text{'s comp} = 1\ 0\ 0\ 1
$$

$\Rightarrow 2\text{'s comp } (+1) =$

$$
\begin{array}{ccccc}
1 & & 1 & & \\
1 & 0 & 1 & 1 & \\
+ & 1 & 0 & 1 & 0 \\
\hline
1 & 0 & 1 & 0 & 1
\end{array}
$$

$\downarrow$

Drop last carry

The main difference between these two methods of subtraction is that with the 1's-complement method the subtrahend is complemented, addition is performed, and then the last carry is added to form the final answer, whereas with the 2's-complement method, the subtrahend is complemented, a 1 is added right away, and then addition is performed, dropping the last carry to get the final answer.

## 8-1g  SERIAL ADDER/SUBTRACTER

Two registers, a full-adder, an exclusive-OR gate, and two D-type flip-flops can be connected together to produce a circuit capable of performing adding and subtracting operations as shown in Figure 8-3. This serial operation of adding and subtracting may not be as practical as parallel addition and subtraction to be covered in the next section, but it provides an ideal understanding of a basic digital system.

Register $A$ is used to contain the augend for adding and to receive the sum answer; also, to contain the minuend for subtraction and to receive the difference answer. Register $B$ is used to contain the addend for adding and the subtrahend for subtracting. The exclusive-OR gate, $EOR_1$, provides a 1's-complement output from register $B$ when subtracting. The $S_o$ output of the full-adder is brought back to the serial input of register $A$. The $C_o$ output of the full-adder is connected to the carry flip-flop, $FF_B$. Flip-flop $FF_A$ is connected to $FF_B$ to provide a "one-shot" action to turn on the carry flip-flop when a 2's-complement subtraction is desired. The Clk line is used to shift the data in the registers into the adder and to set the carry flip-flop when a carry of 1 appears at the $C_o$ output of the adder. The Clr carry line is used to reset the carry flip-flop after a subtraction operation

**Figure 8-3**  Serial adder/subtracter.

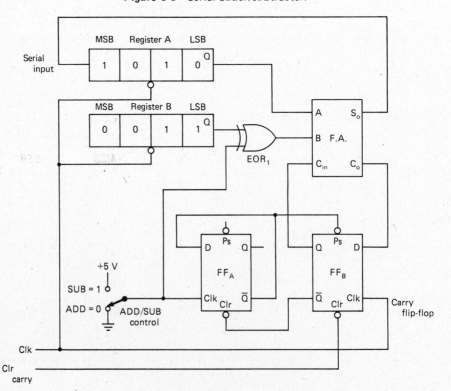

or an overflow indication, where the sum of the number used is larger than 4 bits and cannot be contained in register $A$.

In general, for an add operation the ADD/SUB control switch is at ground or logic 0. This allows the data in register $B$ to be shifted uncomplemented to input $B$ of the adder. The data in both registers are shifted into the adder. As each bit position is added, the $S_o$ output may produce a 1, which is fed back to be shifted into register $A$, or the $C_o$ output may produce a 1, which sets the carry flip-flop. The carry flip-flop provides a 1 at the $C_{in}$ input of the adder on the next shift of data. Four shifts are needed to complete the total add cycle.

A third register could be used to hold the answer for an arithmetic operation, but the method used here reduces the number of components needed. As data are shifted out of register $A$ into the adder, the answer from the adder is being shifted back into register $A$. Most computers and microprocessors operate on a similar principle and this register is referred to as the *accumulator*. In other words, the accumulator temporarily stores the answer to arithmetic operations.

For a subtract operation the ADD/SUB control switch is set to +5 V or a logic 1. This action turns on $FF_A$, which then turns on $FF_B$. As $FF_B$ turns on, its $\overline{Q}$ output goes low, which resets $FF_A$. This is a "one-time" action which places a 1 at the $C_{in}$ input of the adder for a 2's-complement operation. Often, the action is referred to as "set the carry." Because of the 1 on the lower input of $EOR_1$, the data from register $B$ that shift through $EOR_1$ are inverted or 1's complemented. As the data from both registers are shifted through the adder, normal adding is performed, with the difference of the two numbers being placed in register $A$.

The following examples show the contents of the registers, the serial input ($S_o$ of the adder), and the carry flip-flop after each shift for the desired operations. Remember that the LSB of register $A$, the LSB of register $B$ (or the output of $EOR_1$), and the $Q$ output of the carry flip-flop are the data summed in the adder.

**Example 1—The ADD Operation:**

Add these columns

| Shift Pulse | Serial-in ($S_o$) | Register A MSB | | | LSB | Register B MSB | | | LSB | $EOR_1$ out | Carry ($C_{in}$) | |
|---|---|---|---|---|---|---|---|---|---|---|---|---|
| | | | | | | | | | | | 0 | |
| Registers loaded → None | 1 | 1 | 0 | 1 | 0 | 0 | 0 | 1 | 1 | 1 | 0 | |
| 1 | 0 | 1 | 1 | 0 | 1 | 0 | 0 | 0 | 1 | 1 | 1 | ← Carry set |
| 2 | 1 | 0 | 1 | 1 | 0 | 0 | 0 | 0 | 0 | 0 | 0 | |
| 3 | 1 | 1 | 0 | 1 | 1 | 0 | 0 | 0 | 0 | 0 | 0 | |
| 4 | X | 1 | 1 | 0 | 1 | 0 | 0 | 0 | 0 | 0 | 0 | |

Sum

To add:

1. Load registers.
2. Add LSB of register $A$ with the LSB of register $B$ and the carry.
3. Place a 1 at Serial-in if $S_o$ is 1.
4. Place a 1 at Carry if $C_o$ is 1.
5. Shift right one place.
6. Repeat steps 2 through 5 three more times.

*Note:* X means that the value does not matter; the operation is finished.

### Example 2—The SUB Operation:

Add these columns

| Shift Pulse | Serial-in ($S_o$) | Register A | | | | Register B | | | | $EOR_1$ out | Carry ($C_{in}$) | |
|---|---|---|---|---|---|---|---|---|---|---|---|---|
| | | MSB | | | LSB | MSB | | | LSB | | | |
| | | | | | | | | | | | 1 | ← Carry set |
| Registers loaded → None | 1 | 1 | 0 | 1 | 0 | 0 | 0 | 1 | 1 | 0 | 0 | |
| 1 | 1 | 1 | 1 | 0 | 1 | 0 | 0 | 0 | 1 | 0 | 0 | |
| 2 | 1 | 1 | 1 | 1 | 0 | 0 | 0 | 0 | 0 | 1 | 0 | |
| 3 | 0 | 1 | 1 | 1 | 1 | 0 | 0 | 0 | 0 | 1 | 1 | ← Carry set |
| 4 | X | 0 | 1 | 1 | 1 | 0 | 0 | 0 | 0 | 1 | 1 | ← Carry set |

Difference

To subtract:

1. Load registers.
2. Set the carry with a 1.
3. Add LSB of register $A$ with the complemented output of $EOR_1$ from LSB of register $B$ and the carry.
4. Place a 1 at Serial-in if $S_o$ is 1.
5. Place a 1 at Carry if $C_o$ is 1.
6. Shift right one place.
7. Repeat steps 3 through 6 three more times.

*Note:* X means that the value does not matter; the operation is finished.

## 8-1h  PARALLEL ADDER/SUBTRACTER

The serial adder/subtracter requires time to shift the data to the adder for a result, whereas a parallel adder/subtracter, shown in Figure 8-4a, will produce a result nearly instantly. The parallel adder/subtracter requires a full-adder for each bit in the digital word. Since 4-bit word registers are used,

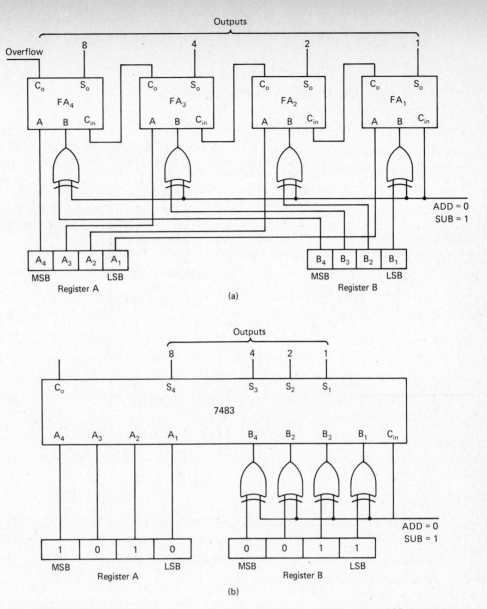

**Figure 8-4** Parallel adder/subtracter: (a) with ripple carry; (b) typical 7483 4-bit binary adder DIP IC.

four full-adders are required. The adders are arranged in ascending order $FA_1$ to $FA_4$. Each LSB of the registers is connected to $FA_1$ ($A_1$ and $B_1$ to $FA_1$). The next-highest-order bits are connected to each adder accordingly until $A_4$ and $B_4$ are connected to $FA_4$. The outputs of register $B$ pass through exclusive-OR gates in order to complement the data for subtraction operations. The same ADD/SUB logic control is used as with the serial adder/subtracter and a line going to $C_{in}$ of $FA_1$ accomplishes 2's-complement subtraction. The $C_o$ of lower-order adders are connected to the $C_{in}$ of higher-order adders and is referred to as *ripple carry*. Outputs 8, 4, 2, and 1 can be connected to another register or fed back to register $A$ to store the result of an operation temporarily. An obvious disadvantage of a parallel adder/subtracter is that more circuits are required.

Several full-adders are produced in IC form. The 7483 4-bit binary full-adder is shown in Figure 8-4b with register connections. If an 8-bit adder is needed, the $C_o$ of one adder is connected to the $C_{in}$ of the next-highest-order adder and the extended registers are connected to the inputs accordingly. More experience will be gained with a parallel adder/subtracter in Experiment 3.

## 8-1i ASSIGNED NUMBERS AND 2's-COMPLEMENT NOTATION

So far we have dealt with positive numbers for binary addition and subtraction. However, some digital circuits, and particularly microprocessors, must operate with positive and negative numbers. One of the best methods devised and currently used in many microprocessors is the 2's-complement representation or notation of numbers. Since some popular microprocessors use either 8-bit or 16-bit word lengths, the 8-bit word will be used here to familiarize you with this type of mathematical operations.

The MSB of a word is used to indicate if the number is positive or negative. If the number is positive, this bit is 0, and if the number is negative, this bit is a 1. The following example shows how a +5 and a −5 would appear in binary form when set in a register.

$$+5_{10} = \text{MSB } 0 \; 0 \; 0 \; 0 \; 0 \; 1 \; 0 \; 1 \text{ LSB}$$

Sign bit — Absolute value of number

$$-5_{10} = \text{MSB } 1 \; 1 \; 1 \; 1 \; 1 \; 0 \; 1 \; 1 \text{ LSB}$$

Sign bit — 2's complement of number

Because the MSB is used to indicate the sign of the number, there remain only 7 bits to express the absolute value of a number. Since a 0 uses one of the positive sign numbers, the maximum positive number that can be contained by an 8-bit register is $+127_{10}$ ($01111111_2 = +127_{10}$). However, the maximum negative number is $-128_{10}$ ($10000000 = -128_{10}$). Therefore, the range of decimal numbers for an 8-bit register is +127 to −128.

The 2's-complement notation of positive numbers appears the same as normal binary numbers, but negative numbers will appear a little strange at first, because they are represented in 2's-complement form. For example, the $-5_{10}$ is first converted to its binary equivalent of $00000101_2$. Next, the 1's complement is found by changing all 0s to 1s, and vice versa; 11111010. Finally, a 1 is added to this to get the 2's-complement number; 11111010 + 1 = $11111011_{\text{2's comp}}$. The following list shows 2's-complement notation for the decimal numbers +9 to −9.

| Decimal | 2's Complement | Decimal | 2's Complement |
|---------|----------------|---------|----------------|
| 0       | 00000000       | —       | —              |
| +1      | 00000001       | −1      | 11111111       |
| +2      | 00000010       | −2      | 11111110       |
| +3      | 00000011       | −3      | 11111101       |
| +4      | 00000100       | −4      | 11111100       |
| +5      | 00000101       | −5      | 11111011       |
| +6      | 00000110       | −6      | 11111010       |
| +7      | 00000111       | −7      | 11111001       |
| +8      | 00001000       | −8      | 11111000       |
| +9      | 00001001       | −9      | 11110111       |

The following examples show how the addition of assigned numbers using 2's-complement notation will appear in a register.

1. *Decimal*     *2's complement*

```
      +5           0000 0101
(+)  +3        +   0000 0011
     ─────          ─────────
      +8           0000 1000
```

2. *Decimal*     *2's complement*

```
      +5           0000 0101
(+)  −3        +   1111 1101
     ─────        ───────────
      +2         1 0000 0010
                       ↓
                  Drop carry
```

3. *Decimal*     *2's complement*

```
      −5           1111 1011
(+)  +3        +   0000 0011
     ─────         ──────────
      −2           1111 1110
```

4. *Decimal*     *2's complement*

```
      −5           1111 1011
(+)  −3        +   1111 1101
     ─────        ───────────
      −8         1 1111 1000
                       ↓
                  Drop carry
```

Remember that in subtraction with assigned numbers, the subtrahend is always 2's complemented and then added. The shortcut to performing this operation is to notice the sign of the subtrahend. For instance, a positive-sign number would appear in negated 2's-complemented form and then added to the minuend, whereas a negative-sign number would appear as a positive number and then added to the minuend. As an example:

$$(-) +3 = 1111 \ 1101 \quad \text{and} \quad (-) -3 = 0000 \ 0011$$

The following examples will show how the subtraction of assigned numbers using 2's-complement notation will appear in a register.

1. *Decimal*     *2's complement*

```
      +5           0000 0101
(−)  +3        +   1111 1101
     ─────        ───────────
      +2         1 0000 0010
                       ↓
                  Drop carry
```

2. *Decimal*     *2's complement*

```
      +5           0000 0101
(−)  −3        +   0000 0011
     ─────         ──────────
      +8           0000 1000
```

3. *Decimal*     *2's complement*

```
      +3           0000 0011
(−)  −5        +   0000 0101
     ─────         ──────────
      +8           0000 1000
```

4. *Decimal*     *2's complement*

```
      +3           0000 0011
(−)  +5        +   1111 1011
     ─────         ──────────
      −2           1111 1110
```

5. *Decimal*     *2's complement*

```
      −5           1111 1011
(−)  +3        +   1111 1101
     ─────        ───────────
      −8         1 1111 1000
                       ↓
                  Drop carry
```

6. *Decimal*     *2's complement*

```
      −5           1111 1011
(−)  −3        +   0000 0011
     ─────         ──────────
      −2           1111 1110
```

7. *Decimal*          *2's complement*

$$
\begin{array}{r}
-3 \\
(-)\ +5 \\
\hline
-8
\end{array}
\qquad
\begin{array}{r}
1111\ 1101 \\
+\quad 1111\ 1011 \\
\hline
1\ 1111\ 1000 \\
\downarrow \\
\text{Drop carry}
\end{array}
$$

8. *Decimal*          *2's complement*

$$
\begin{array}{r}
-3 \\
(-)\ -5 \\
\hline
+2
\end{array}
\qquad
\begin{array}{r}
1111\ 1101 \\
+\quad 0000\ 0101 \\
\hline
1\ 0000\ 0010 \\
\downarrow \\
\text{Drop carry}
\end{array}
$$

Many computers and microprocessors use a hexadecimal number system keyboard for data entry into the system; therefore, it is necessary to convert from binary to hexadecimal and realize how to enter positive and negative numbers. The following two examples will show how this is accomplished.

**Example 1:**

*Decimal*

$$
\begin{array}{r}
+8 \\
(+)\ -3 \\
\hline
+5
\end{array}
$$

*2's complement*

$$
\begin{array}{r}
0000\ 1000 \\
+\quad 1111\ 1101 \\
\hline
1\ 0000\ 0101 \\
\downarrow \\
\text{Drop carry}
\end{array}
$$

*Hexadecimal*

$$
\begin{array}{r}
0\ \ 8 \\
+\ F\ \ D \\
\hline
0\ \ 5
\end{array}
$$

**Example 2:**

*Decimal*

$$
\begin{array}{r}
-5 \\
(-)\ +4 \\
\hline
-9
\end{array}
$$

*2's complement*

$$
\begin{array}{r}
1111\ 1011 \\
+\quad 1111\ 1100 \\
\hline
1\ 1111\ 0111 \\
\downarrow \\
\text{Drop carry}
\end{array}
$$

*Hexadecimal*

$$
\begin{array}{r}
F\ \ B \\
+\ F\ \ C \\
\hline
F\ \ 7
\end{array}
$$

**SECTION 8-2**
**EXERCISES I**

Perform all the exercises in this section before beginning the next section. For exercises involving binary mathematics, you can check your work by converting the binary numbers to decimal numbers and then perform the operation.

1. Add the following binary numbers.

a.
$$
\begin{array}{r}
0101 \\
+\ 0011 \\
\hline
\end{array}
$$

b.
$$
\begin{array}{r}
0010 \\
+\ 1010 \\
\hline
\end{array}
$$

c.
$$
\begin{array}{r}
1001 \\
+\ 0101 \\
\hline
\end{array}
$$

d.
$$
\begin{array}{r}
0011 \\
+\ 1011 \\
\hline
\end{array}
$$

e.
$$
\begin{array}{r}
0111 \\
+\ 0001 \\
\hline
\end{array}
$$

2. Subtract the following binary numbers.

a.
$$
\begin{array}{r}
1011 \\
-\ 0010 \\
\hline
\end{array}
$$

b.
$$
\begin{array}{r}
1101 \\
-\ 0101 \\
\hline
\end{array}
$$

c.
$$
\begin{array}{r}
1001 \\
-\ 0010 \\
\hline
\end{array}
$$

d.
$$
\begin{array}{r}
1110 \\
-\ 0101 \\
\hline
\end{array}
$$

e.
$$
\begin{array}{r}
1000 \\
-\ 0001 \\
\hline
\end{array}
$$

3. Show the following binary numbers in their 1's- and 2's-complement forms.

|  | Number | | 1's complement | | 2's complement |
|---|---|---|---|---|---|
| a. | 0010 | = | __ __ __ __ | = | __ __ __ __ |
| b. | 0101 | = | __ __ __ __ | = | __ __ __ __ |
| c. | 0100 | = | __ __ __ __ | = | __ __ __ __ |
| d. | 1000 | = | __ __ __ __ | = | __ __ __ __ |
| e. | 1101 | = | __ __ __ __ | = | __ __ __ __ |

4. Subtract the following binary numbers using the 1's-complement method of subtraction.

   a.    1000            b.    1010            c.    1111  
       − 0101              − 0111             − 1000

5. Subtract the following binary numbers using the 2's-complement method of subtraction.

   a.    1000            b.    1010            c.    1111  
       − 0101              − 0111             − 1000

6. Draw a simple logic gate diagram of a half-adder.

7. Draw the logic symbol for a half-adder.

8. Draw the truth table for a half-adder and write the Boolean expression for the $S$ and $C$ outputs.

9. Draw the logic gate diagram for a full-adder.

**10.** Draw the logic symbol for a full-adder.

**11.** Draw the truth table for a full-adder and write the Boolean expression for the $S_o$ and $C_o$ outputs.

**12.** Referring to Figure 8-3 and Example 1 in Section 8-1g, add the numbers given in the serial adder and show the contents of the registers, adder, and other components listed after each add and shift.

| Shift Pulse | Serial-in $(S_o)$ | Register A | | | | | Register B | | | | | EOR out | Carry $(C_{in})$ |
|---|---|---|---|---|---|---|---|---|---|---|---|---|---|
| | | MSB | | | LSB | | MSB | | | LSB | | | 0 |
| Registers loaded ⟶ None | ═ | 1 | 0 | 0 | 1 | | 0 | 0 | 1 | 1 | | 1 | — |
| 1 | — | — | — | — | — | | — | — | — | — | | — | — |
| 2 | — | — | — | — | — | | — | — | — | — | | — | — |
| 3 | — | — | — | — | — | | — | — | — | — | | — | — |
| 4 | — | — | — | — | — | | — | — | — | — | | — | — |

**13.** Referring to Figure 8-3 and Example 2 in Section 8-1g, subtract the numbers given in the serial adder and show the contents of the registers, adder, and other components listed after each add and shift.

| Shift Pulse | Serial-in $(S_o)$ | Register A | | | | | Register B | | | | | EOR out | Carry $(C_{in})$ |
|---|---|---|---|---|---|---|---|---|---|---|---|---|---|
| | | MSB | | | LSB | | MSB | | | LSB | | | 1 |
| Registers loaded ⟶ None | — | 1 | 0 | 0 | 1 | | 0 | 0 | 1 | 1 | | 0 | — |
| 1 | — | — | — | — | — | | — | — | — | — | | — | — |
| 2 | — | — | — | — | — | | — | — | — | — | | — | — |
| 3 | — | — | — | — | — | | — | — | — | — | | — | — |
| 4 | — | — | — | — | — | | — | — | — | — | | — | — |

14. List the following phrases under the column that describes the advantages and disadvantages for each method of addition:

   1. Requires the most components
   2. Has the fewest components
   3. Requires the most time to complete an operation
   4. Takes less time to complete an operation

   *Serial Addition*                              *Parallel Addition*


## SECTION 8-3
## DEFINITION EXERCISES

Give a brief description of each of the following terms.

1. Binary addition rules (show examples) _____

   _____

   _____

2. Carry_____

   _____

   _____

3. Binary subtraction rules (show examples) _____

   _____

   _____

4. Borrow_____

   _____

   _____

5. 1's-complement subtraction _____

   _____

   _____

6.  2's-complement subtraction _____

_____

_____

7.  Half-adder _____

_____

_____

8.  Full-adder _____

_____

_____

9.  Serial adder _____

_____

_____

10. Parallel adder _____

_____

_____

11. Representation of assigned numbers: (a) Positive numbers; (b) Negative numbers _____

_____

_____

12. 2's-complement notation arithmetic _____

_____

_____

13. Carry flip-flop _____

_____

_____

14. Set the carry _____

_____

_____

**15.** Clear the carry_____

_____

_____

**SECTION 8-4
EXPERIMENTS**

**EXPERIMENT 1.  BINARY ADDERS**

*Objective:*

To show the operation of a binary half-adder and a binary full-adder.

*Introduction:*

This experiment shows the limitations of a half-adder that can add only 2 bits and a comparison of a full-adder that has a carry input and can add 3 bits.

*Materials Needed:*

*1*  Digital logic trainer *or*

*1*  7486 quad two-input exclusive-OR gate DIP IC

*1*  7408 quad two-input AND gate DIP IC

*1*  7432 quad two-input OR gate DIP IC

*2*  LED indicators

Several hookup wires or leads

*Procedure:*

1.  Construct the circuit shown in Figure 8-5a.

**Figure 8-5**  Binary adders: (a) half-adder; (b) truth table; (c) full-adder; (d) truth table.

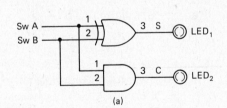

| A | B | S | C |
|---|---|---|---|
| 0 | 0 |   |   |
| 0 | 1 |   |   |
| 1 | 0 |   |   |
| 1 | 1 |   |   |

(a)                                          (b)

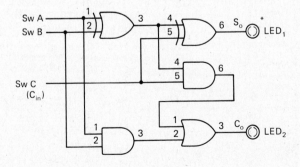

| A | B | $C_{in}$ | $S_o$ | $C_o$ |
|---|---|---|---|---|
| 0 | 0 | 0 |   |   |
| 0 | 0 | 1 |   |   |
| 0 | 1 | 0 |   |   |
| 0 | 1 | 1 |   |   |
| 1 | 0 | 0 |   |   |
| 1 | 0 | 1 |   |   |
| 1 | 1 | 0 |   |   |
| 1 | 1 | 1 |   |   |

(c)                                          (d)

*Note*:  On all ICs, pin 14 = $+V_{CC}$ and pin 7 = Gnd.

2. Set the input switches $A$ and $B$ according to the truth table in Figure 8-5b and record the outputs $S$ and $C$ in their proper place. An LED that is on equals 1 and an LED that is off equals 0.
3. Construct the circuit shown in Figure 8-5c.
4. Set the input switches $A$, $B$, and $C$ according to the truth table in Figure 8-5d and record the outputs, $S_o$ and $C_o$, in their proper place.

*Fill-In Questions:*

1. For a half-adder, when one input is 1 and the other is 0, the $S$ output equals _____ and the $C$ output equals _____.

2. For a half-adder, when both inputs are 1, the $S$ output equals_____ _____ and the $C$ output equals _____.

3. For a full-adder, when only one input is 1, the $S_o$ output equals_____ _____and the $C_o$ output equals_____.

4. For a full-adder, when any two inputs are 1, the $S_o$ output equals_____ _____ and the $C_o$ output equals _____.

5. For a full-adder, when all three inputs are 1, the $S_o$ output equals_____ _____ and the $C_o$ output equals _____.

## EXPERIMENT 2. SERIAL ADDER/SUBTRACTER

*Objective:*

To demonstrate the operation of a serial adder/subtracter.

*Introduction:*

The serial adder/subtracter used in this experiment has one full-adder and the data in registers $A$ and $B$ are shifted into the adder, with the answer or result being shifted back into register $A$. The 7483 IC is a 4-bit binary full-adder, but only one-fourth of the inputs are being used. One exclusive-OR gate in the 7486 IC is used to complement register $B$. The D-type flip-flops in the 7474 IC are used to hold the carry for arithmetic operations. To minimize the number of switches used in this experiment, all data inputs of the registers are connected to ground (0). To set a desired data input to a 1 when loading the registers, simply remove the lead to that input from ground. But remember, once the registers are loaded, connect the data input back to ground or faulty operation will occur when shifting is begun.

*Materials Needed:*

1  Digital logic trainer *or*
2  7495 4-bit shift right/shift left register DIP ICs
1  7486 quad two-input exclusive-OR gate DIP IC
1  7474 dual D-type flip-flop DIP IC

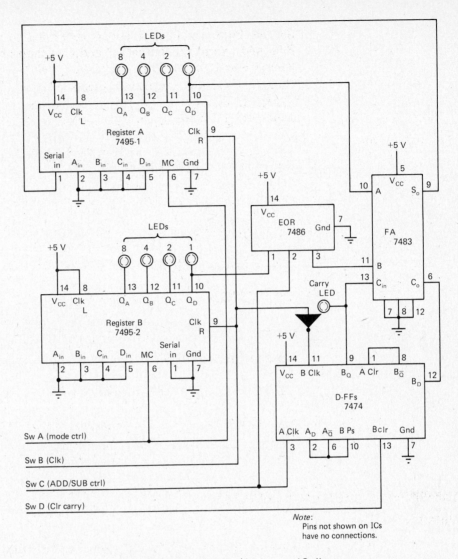

**Figure 8-6** Serial adder/subtracter IC diagram.

*1* 7483 4-bit binary full-adder DIP IC
*9* LED indicators
Several hookup wires or leads

*Procedure:*

### ADD Operation

1. Construct the circuit shown in Figure 8-6.
2. Place all switches ($A$ through $D$) to ground or 0.
3. On register $A$, disconnect data input lead $A_{in}$ (pin 2) and $C_{in}$ (pin 4).
4. On register $B$, disconnect data input lead $C_{in}$ (pin 4) and $D_{in}$ (pin 5).
5. Move Sw $A$ (mode ctrl) up or to a 1.
6. Move Sw $B$ (Clk) up and then down. Register $A$ should contain the data MSB 1010 LSB and register $B$ should contain the data MSB 0011 LSB.
7. Connect all data inputs of registers $A$ and $B$ back to ground.

8. Move Sw $A$ (mode ctrl) down or to a 0.

9. Move Sw $D$ (Clr carry) up and then down. The carry flip-flop should now be reset.

10. Slowly move Sw $B$ (Clk) up and then down and observe the contents of each register. After each shift compare the data in the registers with Example 1 of Section 8-1g. Notice when the carry LED turns on and off. Four shifts will complete this ADD operation. Register $A$ should contain the sum, MSB 1101 LSB, and register $B$ should contain MSB 0000 LSB. The carry LED should be off at this time.

11. Try adding various numbers with the similar procedures as given in steps 2 through 10.

### SUB Operation

12. Place all switches ($A$ through $D$) to ground or 0.

13. On register $A$, disconnect data input lead $A_{in}$ (pin 2) and $C_{in}$ (pin 4).

14. On register $B$, disconnect data input lead $C_{in}$ (pin 4) and $D_{in}$ (pin 5).

15. Move Sw $A$ (mode ctrl) up or to a 1.

16. Move Sw $B$ (Clk) up and then down. Both registers should contain the same data as in step 6.

17. Connect all data inputs of both registers back to ground.

18. Move Sw $A$ (mode ctrl) down or to a 0.

19. Move Sw $C$ (ADD/SUB ctrl) up or to a 1. The carry LED should turn on.

20. Slowly move Sw $B$ (Clk) up and down and observe the contents of each register. After each shift, compare the data in the registers with Example 2 of Section 8-1g. Notice when the carry LED turns on and off. Four shifts will complete this SUB operation. Register $A$ should contain the difference, MSB 0111 LSB, and register $B$ should contain MSB 0000 LSB. The carry LED should be on at this time.

21. Try subtracting various numbers with the similar procedures as given in steps 12 through 20.

*Fill-In Questions:*

1. A 4-bit serial adder/subtracter requires _____ shift pulses to complete an operation.

2. When the ADD/SUB ctrl switch is moved to 1, the _____ flip-flop is set and the output of the exclusive-OR gate will be the _____ _____ of the contents of register $B$.

3. This serial adder/subtracter uses the _____ method of subtraction.

4. The answer coming from output $S_o$ of the full-adder is fed back to _____ _____ .

## EXPERIMENT 3. PARALLEL ADDER/SUBTRACTER

*Objective:*

To show the operation of a parallel adder/subtracter circuit.

*Introduction:*

In this experiment, total use is made of the 7483 binary full-adder IC. Logic switches are used to represent registers $A$ and $B$ and the LEDs show the results of the operations. The carry-in is listed as $C_o$ and the carry-out is listed as $C_4$ for this particular IC. Manufacturers may vary on the designation of terms. The outputs are represented as $\Sigma_1$, $\Sigma_2$, $\Sigma_3$, and $\Sigma_4$. The 7486 exclusive-OR IC is used for 2's-complement subtraction with the adder.

*Materials Needed:*

1   Digital logic trainer or

1   7483 4-bit binary full-adder DIP IC

1   7486 quad two-input exclusive-OR gate DIP IC

5   LED indicators

    Several hookup wires or leads

*Procedure:*

1. Construct the circuit shown in Figure 8-7.
2. Set the switches accordingly: Sw $A = 1$, Sw $B = 0$, Sw $C = 1$, Sw $D = 0$, Sw $E = 0$, Sw $F = 0$, Sw $G = 1$, Sw $H = 1$, and Sw $I = 0$. Register $A$ contains the number MSB 1010 LSB, register $B$ contains the number MSB 0011 LSB, and the circuit is in the ADD operation.
3. Observe the LEDs and indicate which ones are on and off: $\text{LED}_{\text{carry}} =$

_____, $\text{LED}_8 =$ _____, $\text{LED}_4 =$ _____,

$\text{LED}_2 =$ _____, and $\text{LED}_1 =$ _____.

**Figure 8-7** Parallel adder/ subtracter.

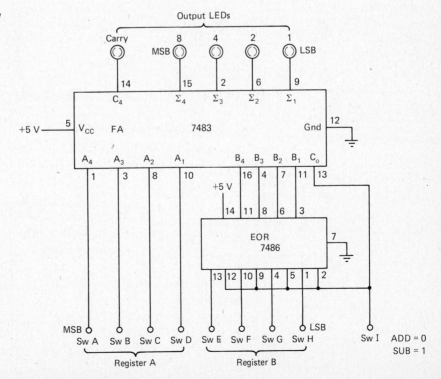

4. Move $Sw_I$ up or to a 1. The circuit is now in the SUB operation.

5. Observe the LEDs and indicate which ones are on and off: $LED_{carry} =$

_____, $LED_8 =$ _____, $LED_4 =$ _____,

$LED_2 =$ _____, and $LED_1 =$ _____.

6. Try adding and subtracting other binary numbers using the procedures similar to steps 2 through 5.

*Fill-In Questions:*

1. The parallel adder performs operations in _____ time than the

serial adder.

2. The output of the exclusive-OR gates when the ADD/SUB control line

equals 1 will be the _____ of register $B$.

**SECTION 8-5
INSTANT REVIEW**

- The basic rules of binary addition are:

$$
\begin{array}{llll}
1. \quad \begin{array}{r} 0 \\ +\,0 \\ \hline 0 \end{array} &
2. \quad \begin{array}{r} 0 \\ +\,1 \\ \hline 1 \end{array} &
3. \quad \begin{array}{r} 1 \\ +\,0 \\ \hline 1 \end{array} &
4. \quad \text{Carry} \begin{array}{r} 1 \\ +\,1 \\ \hline 1\,0 \end{array} \text{(with a carry of 1)}
\end{array}
$$

- The basic rules of binary subtraction are:

$$
\begin{array}{llll}
1. \quad \begin{array}{r} 0 \\ -\,0 \\ \hline 0 \end{array} &
2. \quad \begin{array}{r} 1 \\ -\,0 \\ \hline 1 \end{array} &
3. \quad \begin{array}{r} 1 \\ -\,1 \\ \hline 0 \end{array} &
4. \quad \begin{array}{r} 0 \\ -\,1 \\ \hline 1 \end{array} \text{(with a borrow of 1)}
\end{array}
$$

- A binary half-adder is limited to adding only 2 bits and has no provision for a carry in. It has a sum ($S$) and carry ($C$) output.
- A binary full-adder can add 2 bits and has a carry input. It has a sum-out ($S_o$) and a carry-out ($C_o$) output.
- With the 1's-complement subtraction method the carry must be added in to obtain the correct answer.
- With the 2's-complement subtraction method a 1 is automatically added to the 1's complement and the carry is dropped to obtain the correct answer.
- Serial addition/subtraction requires that each bit be shifted into the adder. It has fewer components, but requires more time to complete an operation.
- Parallel addition/subtraction is nearly instantaneous in operation, but requires more components.
- To identify assigned numbers in computers and microprocessors the MSB

of the binary word is 0 for positive numbers and the MSB of the binary
word is 1 for negative numbers.

- One of the easiest and most efficient ways to perform mathematical
  operations in computers and microprocessors is with the 2's-complement
  notation method.

**SECTION 8-6
EXERCISES II**

Perform all the exercises in this section before beginning the next section.

1. Add the following binary numbers.

   a.    1 0 0 1 0 1 1 1      b.   0 0 0 0 0 0 0 1      c.   0 1 0 0 0 0 0 1
      + 0 1 0 0 1 0 1 1        + 0 1 1 1 1 1 1 1       + 0 1 1 1 1 1 1 1

2. Subtract the following binary numbers.

   a.    1 0 0 1 0 1 1 1      b.   1 0 0 0 0 0 0 0      c.   1 0 0 0 1 0 0 1
      − 0 1 0 0 1 0 1 1        − 0 0 0 0 0 0 0 1       − 0 0 0 1 0 0 1 1

3. Using the 2's-complement notation method, solve the following problems by converting the decimal number to 2's-complement notation and
   then converting to hexadecimal notation.

| | *Decimal Number* | | *2's Complement* | | *Hexadecimal* |
|---|---|---|---|---|---|
| a. | +17 | = | _ _ _ _ _ _ _ _ | = | _ _ |
| | (+) +09 | = | + _ _ _ _ _ _ _ _ | = | + _ _ |
| b. | +36 | = | _ _ _ _ _ _ _ _ | = | _ _ |
| | (+) −24 | = | + _ _ _ _ _ _ _ _ | = | + _ _ |
| c. | +54 | = | _ _ _ _ _ _ _ _ | = | _ _ |
| | (+) −63 | = | + _ _ _ _ _ _ _ _ | = | + _ _ |
| d. | −39 | = | _ _ _ _ _ _ _ _ | = | _ _ |
| | (+) −26 | = | + _ _ _ _ _ _ _ _ | = | + _ _ |

e.     +32        =     _ _ _ _ _ _ _ _      =     _ _
    (−) +18       =   + _____       =   − ==


f.     +44        =     _ _ _ _ _ _ _ _      =     _ _
    (−) −21       =   + _____       =   − ==


g.     +21        =     _ _ _ _ _ _ _ _      =     _ _
    (−) +35       =   + _____       =   − ==


h.     −29        =     _ _ _ _ _ _ _ _      =     _ _
    (−) −37       =   + _____       =   − ==


i.     −23        =     _ _ _ _ _ _ _ _      =     _ _
    (−) +56       =   + _____       =   − ==


j.     −39        =     _ _ _ _ _ _ _ _      =     _ _
    (−) +16       =   + _____       =   − ==


k.     +45        =     _ _ _ _ _ _ _ _      =     _ _
    (−) −73       =   + _____       =   − ==


l.     −99        =     _ _ _ _ _ _ _ _      =     _ _
    (−) −29       =   + _____       =   − ==

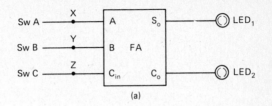

| Condition | A | B | $C_{in}$ | $S_o$ | $C_o$ | Comments |
|---|---|---|---|---|---|---|
| Normal | 0 | 0 | 0 | | | |
| | 0 | 0 | 1 | | | |
| | 0 | 1 | 0 | | | |
| | 0 | 1 | 1 | | | |
| | 1 | 0 | 0 | | | |
| | 1 | 0 | 1 | | | |
| | 1 | 1 | 0 | | | |
| | 1 | 1 | 1 | | | |
| Open point X | 0 | 1 | 0 | | | |
| Open point Y | 1 | 0 | 0 | | | |
| Open point Z | 1 | 1 | 0 | | | |

**Figure 8-8** Troubleshooting a full-adder: (a) logic diagram; (b) troubleshooting chart.

(b)

## SECTION 8-7
## TROUBLESHOOTING APPLICATION: TESTING A BINARY FULL-ADDER

A digital IC can develop internal problems that will keep its outputs at fixed levels, 0 or 1, regardless of the input conditions. However, some conditions at the inputs will cause the same symptoms. An open input will usually appear as a 1 to a digital circuit.

Construct a full-adder as shown in Figure 8-8a from discrete logic gates or one-fourth of a 7483 IC, as used in Experiment 2. Using the troubleshooting chart in Figure 8-8b, set the input conditions $A$, $B$, and $C_{in}$ as indicated and record $S_o$ and $C_o$ in their proper place.

## SECTION 8-8
## SELF-CHECKING QUIZZES

### 8-8a   BINARY ADDERS: TRUE-FALSE QUIZ

Place a T for true or an F for false to the left of each statement.

_____   1.   Adding three binary 1s in a single column results in 1 with a carry of 1.

_____   2.   10010101 minus 01011101 equals 00111000.

_____   3.   A half-adder has three inputs.

_____   4.   The 1's complement of 10011101 is 01100010.

_____ 5. In serial addition, the data are shifted through a single full-adder.

_____ 6. In the 2's-complement method of subtraction the carry is added in.

_____ 7. The 2's complement of 10010000 is 01110000.

_____ 8. When using 2's-complement subtraction the data at input $B$ of the full-adder are actually the 1's complement.

_____ 9. The 2's complement notation of $-29_{10}$ is 11100010.

_____ 10. The 2's-complement notation of $+124_{10}$ is 10000100.

## 8-8b  BINARY ADDERS: MULTIPLE-CHOICE QUIZ

Circle the correct answer for each question.

1. If the inputs to a full-adder are $A = 0$, $B = 1$, and $C_{in} = 1$, the outputs are:

   a. $S_o = 0, C_o = 0$    b. $S_o = 1, C_o = 1$

   c. $S_o = 1, C_o = 0$    d. $S_o = 0, C_o = 1$

2. The binary numbers 10010111 plus 00101110 equals:

   a. 10111111          b. 01000000

   c. 11000001          d. 11000101

3. The binary number 10010111 minus 00101110 equals:

   a. 11000001          b. 11000101

   c. 01101000          d. 01101001

4. If the inputs to a full-adder are $A = 1$, $B = 1$, and $C_{in} = 1$, then the outputs are:

   a. $S_o = 0, C_o = 0$    b. $S_o = 1, C_o = 1$

   c. $S_o = 1, C_o = 0$    d. $S_o = 0, C_o = 1$

5. Using 2's-complement notation, $-56_{10}$ minus $-78_{10}$ equals:

   a. 1 00010110          b. 1 01111010
      ↓                      ↓

   c. 10000110          d. 01111001

6. When using 2's-complement notation, $-23_{10}$ plus $+69_{10}$ will equal in hexadecimal:

   a. 2C          b. D4

   c. 2E          d. 2F

Refer to Figure 8-9 for questions 7 through 10.

7. The carry flip-flop will turn on with:

   a. a 1 at $C_o$ and a clock pulse

   b. ADD/SUB control switch set to SUB

   c. both answers a and b

   d. none of the above

8. The device that accomplishes complementation is:

   a. $FF_A$          b. $FF_B$

   c. $EOR_1$          d. FA

9. For an ADD operation after the second shift, register $A$ contains:

   a. 1101          b. 1011

   c. 0110          d. 1100

10. For a SUB operation, after the third shift, register $A$ contains:

   a. 0110          b. 0011

   c. 1001          d. 1101

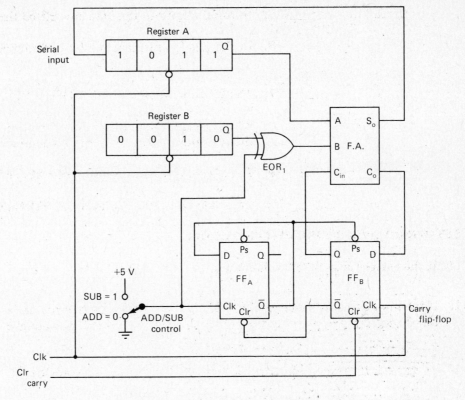

**Figure 8-9**   Circuit for questions 7 through 10.

## ANSWERS TO EXPERIMENTS AND QUIZZES FOR UNIT 8

*Experiment 1.*

(1) 1, 0     (2) 0, 1     (3) 1, 0     (4) 0, 1     (5) 1, 1

*Experiment 2.*

(1) 4     (2) carry, complement     (3) 2's complement     (4) register *A*

*Experiment 3.*

(1) less     (2) complement

*True-False:*

(1) T     (2) T     (3) F     (4) T     (5) T     (6) F     (7) T     (8) T     (9) F
(10) F

*Multiple-Choice:*

(1) d     (2) d     (3) d     (4) b     (5) a     (6) c     (7) c     (8) c     (9) c
(10) b

# Unit 9

---

# Data Conversion
# and Transmission Circuits

## 9-1a  INTRODUCTION

A means is needed to operate or communicate efficiently with digital circuits and computers. Data conversion devices are used to convert or change number systems such as decimal and hexadecimal to the binary number system, and vice versa. Encoders are used to convert data at the input terminals of a digital system such as from a decimal keyboard to binary. Decoders are used within a digital system to execute commands, set up specific operations, and convert from a binary to a decimal output device.

With most normal digital systems, data are transferred between various units or circuits. Data transmission is accomplished efficiently with multiplexers and demultiplexers. Parity circuits are also used to determine if the transmitted data are correct or if a pulse has been lost or added.

## 9-1b  ENCODERS

A typical encoder using NAND gates is shown in Figure 9-1a. The decimal switches are connected to the inputs of the NAND gates. The outputs of the NAND gates are the BCD system. The input switches are normally high (+5 V). This means that a low input is required to cause an action and is referred to as *active-low input*. An input that is required to go high to cause an action is called an *active-high input*. When a particular switch is activated, it goes low (0) and the corresponding outputs of the NAND gates go high and are referred to as *active-high outputs*. For example, when switch 5 is activated, $ND_1$ and $ND_3$ outputs will be high, while $ND_2$ and $ND_4$ outputs will be low.

Encoders can be made from AND, OR, NAND, or NOR gates, a diode

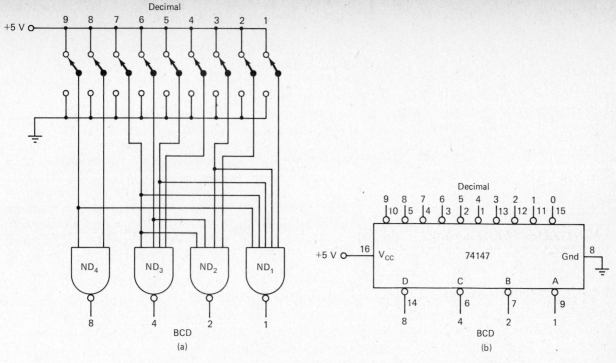

**Figure 9-1** Encoders: (a) simple NAND gate circuit; (b) 74147 DIP IC encoder.

matrix, or special ICs. The 74147 DIP IC encoder is shown in Figure 9-1b. In this case, an active-low input produces active-low outputs. For example, if decimal input 5 is brought low, then outputs $A$ (pin 9) and $C$ (pin 6) would be low and $B$ (pin 7) and $D$ (pin 14) would be high. Many IC devices have active-low outputs in order to preset registers and/or provide a "sink" path for LEDs.

## 9-1c. DECODERS

### 9-1c.1 Basic AND Gate Decoders

Decoders are logical circuits that respond to specific input conditions as shown in Figure 9-2. When the binary up-counter reaches the number 6, AND gate $AN_1$ operates and the LED turns on. The input conditions for $AN_1$ from the flip-flops in the counter are $\overline{A}$, $B$, $C$, and $\overline{D}$. Any other number will not allow $AN_1$ to turn on. As the counter continues to count up, AND gate $AN_2$ will operate when the number 12 is reached and this LED will turn on. The input conditions for $AN_2$ are $\overline{A}$, $\overline{B}$, $C$, and $D$. Any other number will not allow $AN_2$ to turn on.

**Figure 9-2** Basic AND gate decoders.

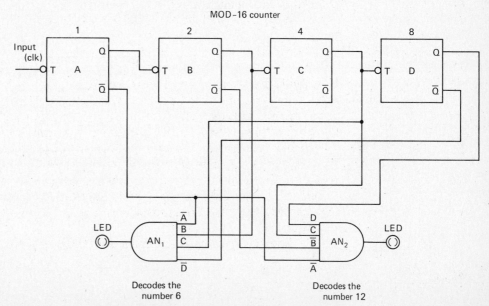

### 9-1c.2 8421 BCD-to-Decimal Decoder

Decoders are used between a register or counter and an output display. A typical 8421 BCD-to-decimal decoder is shown in Figure 9-3a. This decoder accepts 1s at the inputs to produce a corresponding 0 output. For example, if all inputs are 0, then $ND_0$ is off and its output is 0, while all the other NAND gates are on and the rest of the decimal outputs are 1. If the inputs change to $A = 0$, $B = 1$, $C = 1$, and $D = 0$, $ND_6$ turns off and output 6 goes low, while all the other NAND gates are on and their respective outputs are 1. Figure 9-3b shows a typical 7442 one-out-of-ten decoder DIP IC. A truth table is given in Figure 9-3c for both decoders. Input binary numbers above 9 will produce all 1s at the decimal outputs, but this is not shown by the truth table.

**Figure 9-3**  8421 BCD-to-decimal decoder: (a) logic diagram; (b) 7442 DIP IC; (c) truth table.

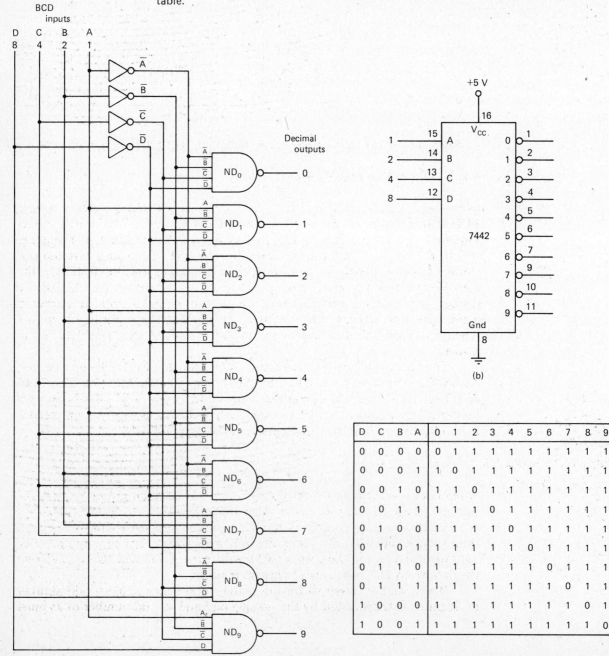

(a)

(b)

(c)

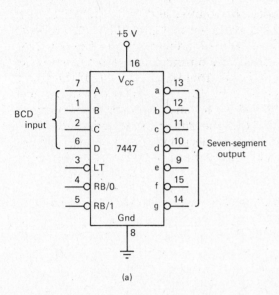

| BCD | | | | Seven-segment | | | | | | | Display |
|---|---|---|---|---|---|---|---|---|---|---|---|
| D | C | B | A | a | b | c | d | e | f | g | Symbol |
| 0 | 0 | 0 | 0 | 0 | 0 | 0 | 0 | 0 | 0 | 1 | 0 |
| 0 | 0 | 0 | 1 | 1 | 0 | 0 | 1 | 1 | 1 | 1 | 1 |
| 0 | 0 | 1 | 0 | 0 | 0 | 1 | 0 | 0 | 1 | 0 | 2 |
| 0 | 0 | 1 | 1 | 0 | 0 | 0 | 0 | 1 | 1 | 0 | 3 |
| 0 | 1 | 0 | 0 | 1 | 0 | 0 | 1 | 1 | 0 | 0 | 4 |
| 0 | 1 | 0 | 1 | 0 | 1 | 0 | 0 | 1 | 0 | 0 | 5 |
| 0 | 1 | 1 | 0 | 1 | 1 | 0 | 0 | 0 | 0 | 0 | 6 |
| 0 | 1 | 1 | 1 | 0 | 0 | 0 | 1 | 1 | 1 | 1 | 7 |
| 1 | 0 | 0 | 0 | 0 | 0 | 0 | 0 | 0 | 0 | 0 | 8 |
| 1 | 0 | 0 | 1 | 0 | 0 | 0 | 0 | 1 | 0 | 0 | 9 |
| 1 | 0 | 1 | 0 | 1 | 1 | 1 | 0 | 0 | 1 | 0 | ⊏ |
| 1 | 0 | 1 | 1 | 1 | 1 | 0 | 0 | 1 | 1 | 0 | ⊐ |
| 1 | 1 | 0 | 0 | 1 | 0 | 1 | 1 | 1 | 0 | 0 | ⊔ |
| 1 | 1 | 0 | 1 | 0 | 1 | 1 | 0 | 1 | 0 | 0 | ⊑ |
| 1 | 1 | 1 | 0 | 1 | 1 | 1 | 0 | 0 | 0 | 0 | ⊢ |
| 1 | 1 | 1 | 1 | 1 | 1 | 1 | 1 | 1 | 1 | 1 | Blank |

(b)

**Figure 9-4** 8421 BCD-to-seven-segment decoder: (a) 7447 DIP IC; (b) truth table.

### 9-1c.3 8421 BCD-to-Seven-Segment Decoder

Seven-segment LED displays are covered in Unit 10. These displays need a BCD-to-seven-segment decoder as shown in Figure 9-4a. The *ABCD* inputs require 1s for proper conversion, while the respective outputs, *a* through *g*, will be 0 to operate the display correctly. Pin 3 (LT) is a lamp test and will cause all seven outputs to go to 0 when it is brought low. Pin 5 (RB/1), the zero blanking input, when low, is used to blank (or turn off all LEDs) the display if it contains a 0. This is useful for turning off insignificant zeros in a multiple-digit display. Pin 4 (RB/0), ripple blanking output, will go low when pin 5 and all data inputs go low and is used to extinguish the entire display.

The truth table shown in Figure 9-4b is for the 7447 DIP IC decoder. Notice the symbols produced with a BCD input greater than nine. The 7446 and 7447 BCD-to-seven-segment decoders have active-low outputs and are used with common-anode-type displays, but the 7448 decoder has active-high outputs and is used with common-cathode-type displays.

### 9-1d  PARITY CIRCUITS

When data are transmitted between units in a system, conditions may occur that can cause the data to lose or gain 1s, thus creating problems. A method of error detection and correction for transmitted data uses parity. *Parity* is a term used to identify whether a data word has an odd or an even number of 1s. For example, the data word 1011 with three 1s has *odd parity*, whereas the data word 1010 with two 1s has *even parity*.

If a system is designed for odd parity, this means that an odd number of 1s must be transmitted by the sending unit and an odd number of 1s must

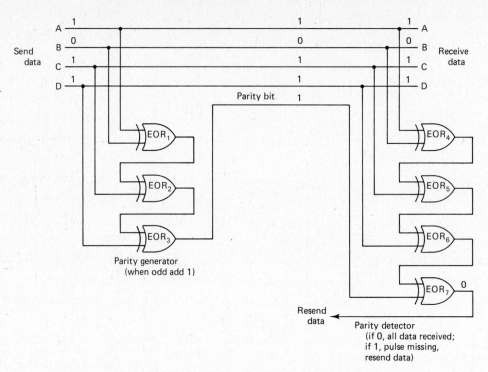

**Figure 9-5** Parity generator/detector system (even parity).

be received by the receiving unit to assume that no error occurred during transmission. If the data word in an odd-parity system has an even number of 1s, an extra 1, known as a *parity bit*, must be generated to make the transmitted data odd. Similarly, for even parity an even number of 1s are sent and received.

Exclusive-OR gates can be used to generate parity bits and detect parity as shown in Figure 9-5. This system uses even parity. The inputs are $A$, $B$, $C$, and $D$ at the send data point. Gates $EOR_1$, $EOR_2$, and $EOR_3$ form a parity generator that monitors the input data. When the input data have an odd number of 1s, $EOR_3$ produces a 1 that is sent over the parity bit line. When input data have an even number of 1s, the output of $EOR_3$ is 0.

Gates $EOR_4$, $EOR_5$, $EOR_6$, and $EOR_7$ form a parity detector that monitors the receive data lines. If the incoming data have an even number of 1s, the output or $EOR_7$ is 0, indicating that the data received are assumed to be correct. When the incoming data have an odd number of 1s, the output of $EOR_7$ is a 1, indicating that the data are incorrect. This 1 is sent back to the data sending unit as a manner of asking to resend the data so that they can be received correctly.

## 9-1e  MULTIPLEXER/DEMULTIPLEXER CIRCUITS

In a digital system, data are transferred directly in parallel through PC boards and wires. However, with the transmission of data over greater distances, wiring can become bulky and expensive. A method is used to convert many parallel data lines to a few lines with serial data. This method is called *multiplexing and demultiplexing*. A basic switch analogy of this system is shown in Figure 9-6a. A basic *multiplexer* (MX) is a switching device capable of selecting one of a number of inputs and connecting it to a single common output. A basic *demultiplexer* (DMX) is the opposite of a multiplexer—a

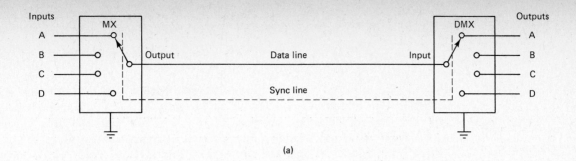

(a)

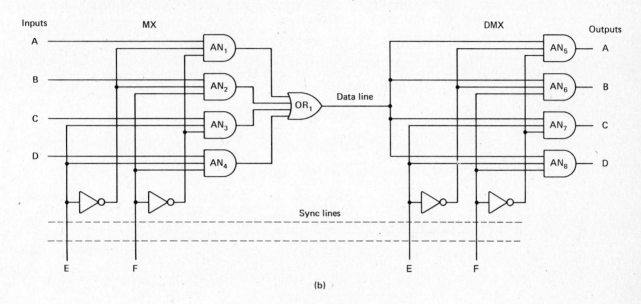

(b)

| Desired input/ output | Control lines | |
|:---:|:---:|:---:|
| | E | F |
| A | 0 | 0 |
| B | 0 | 1 |
| C | 1 | 0 |
| D | 1 | 1 |

(c)

**Figure 9-6** Multiplexer/demultiplexer system: (a) switch analogy; (b) logic gate circuit; (c) truth table.

switching device capable of connecting a single common input to one selected output of a number of outputs. In a basic system the data line is connected between the MX and DMX, but the switching devices must be synchronized (sync) in order for the data to be transmitted properly and received on their respective input/output lines. When the data line is connected to input $A$ of the MX, it must, at the same time, be connected to output $A$ of the DMX. This procedure is the same for inputs $B$, $C$, and $D$ and outputs $B$, $C$, and $D$ to accomplish data transmission and reception in the correct order.

A basic 4-to-1 and 1-to-4 MX/DMX logic gate system is shown in Figure 9-6b. Lines $A$, $B$, $C$, and $D$ are inputs and outputs, respectively. Lines $E$ and $F$ are the control lines for selecting and synchronizing the system. When lines $E$ and $F$ are both 0, gates $AN_1$, $OR_1$, and $AN_5$ are selected to be able to send data from $A$ input to $A$ output. When line $E$ is 0 and line $F$ is 1, gates $AN_2$, $OR_1$, and $AN_6$ are selected to be able to send data from

$B$ input to $B$ output. The truth table shown in Figure 9-6c lists the remaining conditions for control lines $E$ and $F$ to be able to select data lines $C$ and $D$.

The control data appearing at lines $E$ and $F$ are in the form of a sequence, similar to a MOD-4 counter, which causes a synchronized scanning action at the input and output lines $A$, $B$, $C$, and $D$.

A multiplexer is also known as a *data selector* and a demultiplexer is sometimes called a *data distributor*.

**SECTION 9-2**
**EXERCISES I**

Perform all the exercises in this section before beginning the next section.

1. Referring to Figure 9-1a, list which gates turn on for the following decimal numbers of the encoder.

| Decimal Switch | NAND Gates ON (1 at the output) |
|:---:|:---|
| 1 | |
| 2 | |
| 3 | |
| 4 | |
| 5 | |
| 6 | |
| 7 | |
| 8 | |
| 9 | |

2. Draw a MOD-16 counter and decode the numbers $7_{10}$ and $13_{10}$ using AND gates (refer to Figure 9-2 as a guide).

3. Referring to Figure 9-3, list which NAND gates operate and the logic condition at its output for the given BCD inputs.

| BCD Inputs | NAND Gate | NAND Gate Output |
|---|---|---|
| D C B A | That Operates | |
| 0 0 1 1 | | |
| 0 1 1 0 | | |
| 0 1 1 1 | | |
| 1 0 0 1 | | |
| 1 1 0 0 | | |

4. Referring to Figure 9-4, list the logic condition of the seven-segment outputs and the display symbol for the BCD inputs given.

| BCD Inputs | Seven-Segment Outputs | | | | | | | Display Symbol |
|---|---|---|---|---|---|---|---|---|
| D C B A | a | b | c | d | e | f | g | |
| 0 0 1 0 | | | | | | | | |
| 0 1 0 0 | | | | | | | | |
| 0 1 1 1 | | | | | | | | |
| 1 0 1 0 | | | | | | | | |
| 1 1 1 1 | | | | | | | | |

5. List the following numbers as odd parity or even parity.

a. 1 1 0 1 = _____ parity          b. 1 1 1 0 = _____ parity

c. 1 0 0 1 = _____ parity          d. 0 1 1 0 = _____ parity

e. 1 1 0 0 = _____ parity          f. 1 0 1 1 = _____ parity

6. Write YES if you agree or NO if you do not agree with the numbers under each parity heading.

|       | *Even Parity* | *Agree* |       | *Odd Parity* | *Agree* |
|-------|---------------|---------|-------|--------------|---------|
| a.    | 1 1 1 1       | _____ | b.    | 1 0 0 1      | _____ |
| c.    | 1 1 1 0       | _____ | d.    | 0 0 0 1      | _____ |
| e.    | 0 1 0 1       | _____ | f.    | 1 0 1 0      | _____ |
| g.    | 1 0 0 1       | _____ | h.    | 0 1 1 0      | _____ |
| i.    | 1 0 1 1       | _____ | j.    | 1 0 0 0      | _____ |
| k.    | 1 1 1 1       | _____ | l.    | 0 1 1 0      | _____ |

7. Draw a simple multiplexer/demultiplexer system using rotary switches (refer to Figure 9-6a).

8. Referring to Figure 9-6b, list the AND gates that are selected for the following conditions at the $E$ and $F$ control lines.

| Control Lines | | |
|---------------|---|--------------------|
| *E* | *F* | *AND Gates Selected* |
| 0 | 0 | |
| 0 | 1 | |
| 1 | 0 | |
| 1 | 1 | |

## SECTION 9-3
## DEFINITION EXERCISES

Give a brief description of each of the following terms.

1. Encoder_____

_____

_____

2. Decoder _____

_____

_____

3. Odd parity _____

_____

_____

4. Even parity _____

_____

_____

5. Parity generator _____

_____

_____

6. Parity detector _____

_____

_____

7. Multiplexer _____

_____

_____

8. Demultiplexer _____

_____

_____

9. Data selector _____

_____

_____

10. Data distributor _____

_____

_____

**11.** Active-low output _____

_____

_____

**12.** Active-high output _____

_____

_____

**13.** Active-low input _____

_____

_____

**14.** Active-high input _____

_____

_____

**SECTION 9-4
EXPERIMENTS**

**EXPERIMENT 1.  NAND GATE ENCODER**

*Objective:*

To show how NAND gates can be used to convert from the decimal number system to the BCD system.

*Introduction:*

The three ICs used in this experiment form the NAND gate logic circuit shown in Figure 9-1a. Figure 9-7b shows how a five-input NAND gate can be constructed. It is also possible to perform this experiment with the 74147 decimal-to-BCD encoder IC as shown in Figure 9-1b.

*Materials Needed:*

*1*  Digital logic encoder *or*
*1*  7400 quad two-input NAND gate DIP IC
*2*  7420 dual four-input NAND gate DIP ICs
*4*  LED indicators
Several hookup wires or leads

*Procedure:*

1. Construct the circuit shown in Figure 9-7a.
2. Place all switches (Sw 1 to Sw 9) high.
3. Activate each decimal switch one at a time and notice the BCD output LEDs. Indicate the condition of the LEDs in the data table shown in Figure 9-7c. (LED on = 1 and LED off = 0.)

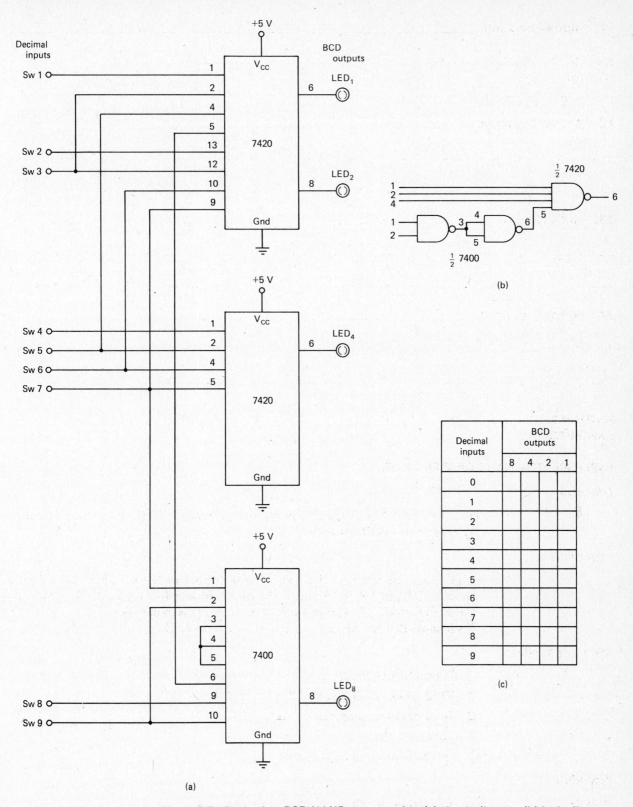

**Figure 9-7** Decimal-to-BCD NAND gate encoder: (a) circuit diagram; (b) logic diagram for constructing a five-input NAND gate; (c) data table.

*Fill-In Questions:*

1.  An encoder can be used to convert from _____ to _____.

2.  With the NAND gate encoder, when Sw 6 is activated, $LED_1$ = _____ _____, $LED_2$ = _____, $LED_3$ = _____, and $LED_4$ = _____.

3.  Some encoders will have an active-low input to produce an active-high output, and some will have an active-_____ input to produce an active-_____ output.

## EXPERIMENT 2.  BASIC AND GATE DECODERS

*Objective:*

To show the basic operation of decoders connected to a counter.

*Introduction:*

This experiment uses a MOD-16 counter and two NAND gates as decoders. Four inverters are used to provide $\overline{Q}$ outputs from the counter, and two inverters at the output of the NAND gates results in AND gate operation. (Refer to Figure 9-2 for the logic circuit.)

*Materials Needed:*

1   Digital logic trainer *or*
1   7404 hex inverter DIP IC
1   7420 dual four-input NAND gate DIP IC
1   7493 4-bit binary counter DIP IC
2   LED indicators
Several hookup wires or leads

*Procedure:*

1.  Construct the circuit as shown in Figure 9-8.
2.  Use Sw *A* to enter pulses into the counter. Notice that when the counter contains the number 6, $LED_6$ will turn on; and similarly for $LED_{12}$ when the counter contains the number 12.
3.  Referring to Figure 9-8b, list the decoded letter inputs on the AND gates for the numbers 3 and 10.
4.  Rewire the 7420 IC for these two new decoders.
5.  Using Sw *A*, enter pulses into the counter. One LED should turn on for the count of 3 and the other LED should turn on for the count of 10.

*Fill-In Questions:*

1.  A decoder can be used to give a control pulse for a specific _____.

2.  To decode the number 11, the inputs to the AND gate decoder from the

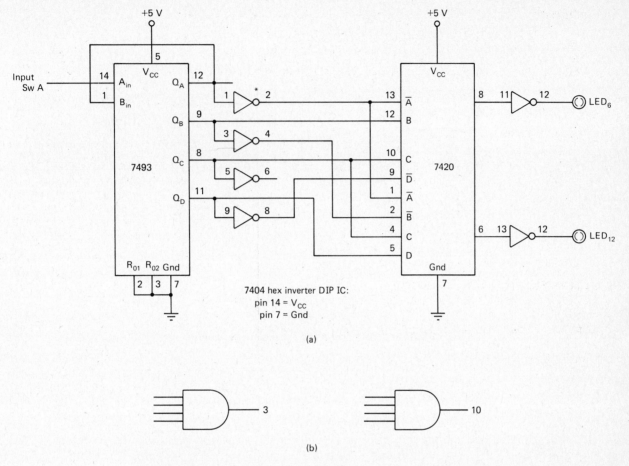

**Figure 9-8**  Basic decoders: (a) MOD-16 counter with NAND gates; (b) AND gate decoded inputs.

MOD-16 counter would be _____, _____, _____,

and _____.


# EXPERIMENT 3.  BCD-TO-DECIMAL DECODER

*Objective:*

To demonstrate the operation of a BCD-to-decimal decoder.

*Introduction:*

This experiment uses four input switches to the IC that control the on/off condition of 10 outputs. A single LED is used to simulate a logic probe to test the condition at each output.

*Materials Needed:*

1   Digital logic trainer *or*
1   7442 BCD-to-decimal decoder DIP IC
1   LED indicator
    Several hookup wires or leads

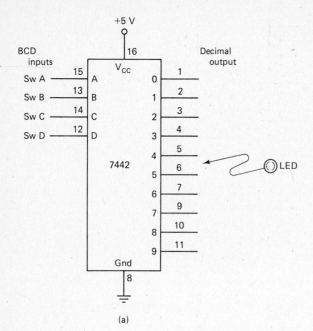

**Figure 9-9**  BCD-to-decimal decoder: (a) circuit diagram; (b) truth table.

*Procedure:*

1. Connect the circuit as shown in Figure 9-9a.
2. Place BCD switches SW *A* through Sw *D* low or at ground.
3. Using the LED as a logic probe, examine each decimal output and list the condition in the truth table shown in Figure 9-9b.
4. Set the input switches for each condition as shown in the truth table and record the output conditions as in step 3.

*Fill-In Questions:*

1. The input numbers to the 7442 IC requires an active-_____ input.

2. The output of the 7442 IC indicates the selected decimal number as an active-_____ output.

3. If the inputs to the 7442 IC are $A = 1$, $B = 0$, $C = 0$, and $D = 1$, the decimal output _____ will be _____.

## EXPERIMENT 4.  PARITY CIRCUITS

*Objective:*

To show the operation of a parity generator and a parity decoder.

*Introduction:*

Exclusive-OR gates are used in this experiment to produce and detect even parity. Refer to Figure 9-5 for the entire system.

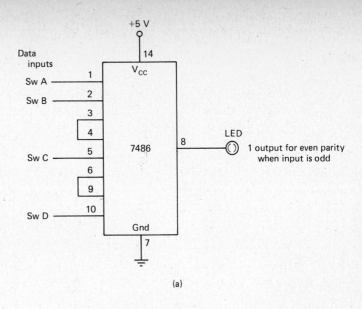

(a)

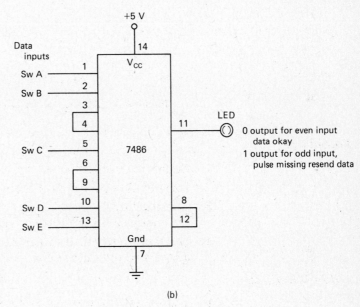

**Figure 9-10** Parity circuits:
(a) parity generator; (b) parity
detector.

(b)

*Materials Needed:*

    *1*   Digital logic trainer *or*

    *1*   7486 quad two-input exclusive-OR gate DIP IC

    *1*   LED indicator

        Several hookup wires or leads

*Procedure:*

    *Parity Generator*

1. Connect the circuit shown in Figure 9-10a.
2. Place all data input switches to 0.
3. Place Sw *A* to a 1. The input data have an odd number of 1s and the output LED is _____.

4. Place Sw *A* and Sw *B* to a 1. The input data have an even number of 1s and the output LED is _____.

5. Try other combinations of odd and even data with the input switches and note the condition of the output.

*Parity Detector*

6. Modify the existing circuit by adding the wiring to pins 8, 12, and 13 as shown in Figure 9-10b.
7. Repeat steps 3, 4, and 5 and note the same output conditions; however, this circuit is used as a detector (see Figure 9-5).

*Fill-In Questions:*

1. For even parity, when the input data contain an even number of 1s, the parity generator's output will be _____.

2. For even parity, when the input data contain an odd number of 1s, the parity generator's output will be _____.

3. The parity detector operates similar to the parity generator; therefore, for even parity when its output is 1, the input data contained an _____ _____ number of 1s.

4. For even parity, if the output of the parity detector is 0, it is assumed that the _____ _____ is okay.

5. For even parity, if a pulse is missing, the parity detector will produce a 1, indicating to _____ the data.

## EXPERIMENT 5.  MULTIPLEXER/DEMULTIPLEXER SYSTEM

*Objective:*

To demonstrate a basic multiplexer/demultiplexer system.

*Introduction:*

Eight data inputs are multiplexed onto a single data line and then demultiplexed to eight data output LEDs. The MX IC is sequenced by a MOD-8 counter and the DMX IC is sequenced by another MOD-8 counter. Both counters are synchronized by the clock. Unselected outputs of the 74155 DMX IC are high; therefore, all of the LEDs will be on. When a channel is selected on both ICs, a low or 0 will be entered at the proper data input and sent over the data line to turn off the corresponding LED.

*Materials Needed:*

1  Digital logic trainer *or*
1  74151 eight-channel digital multiplexer DIP IC
1  74155 dual 2:4 demultiplexer DIP IC

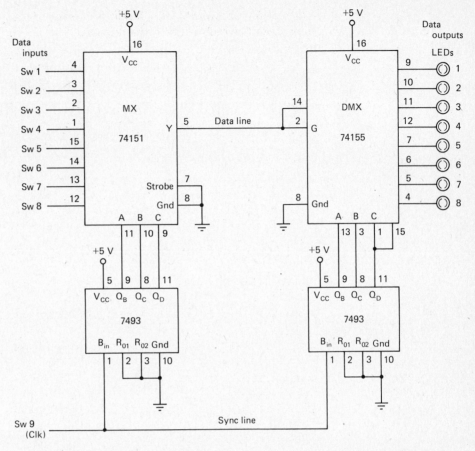

**Figure 9-11**  8-to-1 (1-to-8) bit multiplexer/demultiplexer data transmission system.

2  7493 4-bit binary counter DIP ICs

8  LED indicators

Several hookup wires or leads

*Procedure:*

1.  Connect the circuit shown in Figure 9-11.
2.  Make sure that both counters are reset to 0. Test their outputs with a logic probe or LED indicator.
3.  Place all input switches to a 1. All the LEDs should be on.

4.  Place Sw 1 to a 0. Which data output LED turns off?_____. (Channel 1 is selected.)
5.  Operate Sw 9 (Clk) high and then low. Does the LED remain off?

     _____. (Channel 2 has been selected.)

6.  Move Sw 1 to a 1.

7.  Move Sw 2 to a 0. Does any LED turn off?_____. If yes,

    which one? _____.

8. Operate Sw 9 (Clk) high and low seven times.

9. Place all input switches to a 0. Notice that only LED$_1$ is off.

10. Operate Sw 9 (Clk) and notice the output LEDs. As each channel is selected, if a 0 is present at the data input, the corresponding output LED will turn off. If the counters are "out of sync," the data at the inputs will appear on the wrong channels at the outputs. Experiment with this situation until you fully understand the importance of synchronization.

*Fill-In Questions:*

1. The MX IC converts _____ input line(s) to _____ output line(s).

2. The DMX IC converts _____ input line(s) to _____ output line(s).

3. The counters are used to _____ the operation of the MX and DMX ICs.

4. The counters must have a _____ line between them to ensure the proper transmission of data.

5. For any output LED of the DMX to be able to turn off, it must be connected to the _____ input channel of the MX and that input data line must be a _____.

**SECTION 9-5
INSTANT REVIEW**

- An encoder is a conversion device that is usually used to communicate with a digital system, such as from decimal to BCD.
- A decoder is a conversion device that can execute commands, set up conditions for specific operations, and convert from the binary number system to decimal, seven-segment, or some other number system.
- Odd parity is the transmission of data containing an odd number of 1s. Even parity is the transmission of data containing a even number of 1s.
- A parity generator monitors the outgoing data and may or may not add a 1 to the data, depending on the type of parity being used.
- A parity detector is similar to a parity generator, except that it monitors the incoming data and will produce a 1 that can be used as a signal to ask to resend data if the type of parity is not correct.
- A multiplexer is a synchronized device that connects many input lines to a selected single or few output lines.
- A demultiplexer is a synchronized device that connects a single or a few input lines to many output lines.

**SECTION 9-6**
**EXERCISES II**

Perform all the exercises in this section before beginning the next section.

1. Match the statements in column A to their respective device in column B.

*Column A*                                                              *Column B*

a. This circuit will ask to resend data if a pulse
   is missing.                                          _____ 1. Encoder

b. This device connects many lines to a single
   line.                                                _____ 2. Decoder

c. This circuit is usually used to communicate
   with a digital system.                              _____ 3. Parity generator

                                                        _____ 4. Parity detector
d. This device will transmit a 1 if it is needed.

e. This circuit connects a single input line to        _____ 5. Multiplexer
   many output lines.
                                                        _____ 6. Demultiplexer
f. This device is used to execute commands
   and converts binary numbers to other
   number systems.

2. The data from a register are LSB $A\ B\ C\ D\ E\ F\ G\ H$ MSB. Decode
   the input data to the following AND gate decoders. Show the binary
   number and the hexadecimal number equivalent.

**Example:**

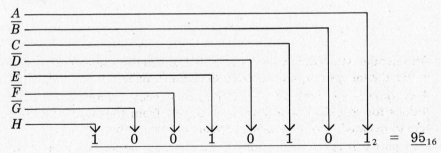

$$1\quad 0\quad 0\quad 1\quad 0\quad 1\quad 0\quad 1_2\ =\ \underline{95}_{16}$$

a.  $A$
    $\overline{B}$
    $\overline{C}$
    $D$
    $\overline{E}$
    $F$
    $\overline{G}$
    $H$

    _____₂ = _____₁₆

b.  $\overline{A}$
    $\overline{B}$
    $C$
    $\overline{D}$
    $E$
    $F$
    $G$
    $H$

    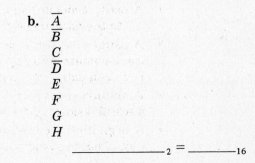

    _____₂ = _____₁₆

c.  $A$
    $\overline{B}$
    $\overline{C}$
    $\overline{D}$
    $\overline{E}$
    $\overline{F}$
    $\overline{G}$
    $H$

    _____$_2$ = _____$_{16}$

d.  $\overline{A}$
    $\overline{B}$
    $\overline{C}$
    $D$
    $E$
    $\overline{F}$
    $G$
    $H$

    _____$_2$ = _____$_{16}$

3. Write YES if you agree or NO if you do not agree with the numbers under each parity heading.

| Even Parity | Agree | | Odd Parity | Agree |
|---|---|---|---|---|
| a. 11010110 | _____ | b. | 10000001 | _____ |
| c. 10011001 | _____ | d. | 11110000 | _____ |
| e. 10010101 | _____ | f. | 11001110 | _____ |

## SECTION 9-7
## TROUBLESHOOTING APPLICATION: TESTING A MULTIPLEXER AND A DEMULTIPLEXER

Troubleshooting and testing digital conversion devices require that you know which input conditions produce the desired output conditions. You already have experience in testing logic gates from Units 2 and 3. You also tested the more complex 7442 BCD-to-decimal decoder IC in Experiment 3 of Section 9-4, by completing its truth table. In testing the multiplexer and demultiplexer of this troubleshooting application, you will have to digitally select the proper input and output to be tested.

For the multiplexer, construct the circuit as shown in Figure 9-12a. Set the select switches as shown in the test chart of Figure 9-12b. Place a logic probe (or LED indicator) on the output. Place a logic pulser (or logic input switch) on the selected input and operate it several times. The output should go high and low. Place a check mark in the output condition column of the test chart. If the output does not respond, place an X in the output condition column. Perform this same operation on all inputs.

The test procedure for the demultiplexer is similar, except that the logic pulser is connected to the input and the logic probe is placed on each output selected.

## SECTION 9-8
## SELF-CHECKING QUIZZES

### 9-8a  DIGITAL CONVERSION DEVICES: TRUE-FALSE QUIZ

Place a T for true or an F for false to the left of each statement.

_____  1.  An encoder can convert from BCD to decimal.

_____  2.  The binary number 11011001 is an example of even parity.

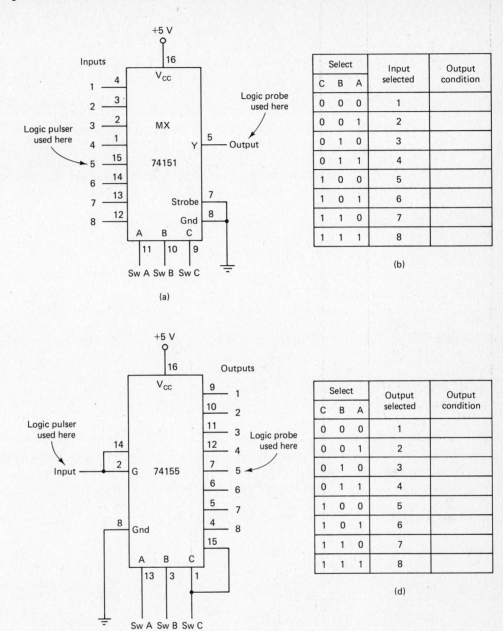

**Figure 9-12**  Testing data transmission ICs: (a) multiplexer; (b) multiplexer test chart; (c) demultiplexer; (d) demultiplexer test chart.

_____ 3. There are four inputs to a BCD-to-decimal decoder.

_____ 4. A multiplexer can be used to switch the data on several lines to a single line.

_____ 5. Decoders can be used to execute commands for specific operations.

_____ 6. An even-parity generator will produce a 1 if the input data are odd.

_____ **7.** The outputs of a demultiplexer are selected digitally with specific inputs.

_____ **8.** Synchronization is not required for a multiplexer/demultiplexer system.

_____ **9.** A decoder is required between a register (or counter) and a display.

_____ **10.** Active-low outputs means that 0s produce the desired action of a device.

## 9-8b  DIGITAL CONVERSION DEVICES: MULTIPLE-CHOICE QUIZ

Circle the correct answer for each question.

**1.** A digital device capable of converting data from serial mode to parallel mode is the:

   **a.** encoder          **b.** decoder

   **c.** multiplexer      **d.** demultiplexer

   **e.** parity generator   **f.** parity detector

**2.** A BCD-to-seven-segment decoder IC has an input that causes all outputs to become active (which results in all segments of the display turning on) and is usually labeled:

   **a.** RB/1           **b.** RB/0

   **c.** LT              **d.** strobe

**3.** An exclusive-OR gate odd-parity detector will produce an output of 1 for the number:

   **a.** 10111100      **b.** 11010011

   **c.** 01101001      **d.** 00011100

**4.** A device that is used between a keyboard and display register in a digital system is called:

   **a.** an encoder      **b.** a decoder

   **c.** a multiplexer    **d.** a demultiplexer

Questions 5 and 6 refer to Figure 9-13.

**5.** To produce a 0 output for the binary number MSB 0101 LSB, the inputs to the NAND gate would be:

   **a.** $A \overline{B} C \overline{D}$       **b.** $\overline{A} B \overline{C} D$

   **c.** $\overline{A} \overline{B} C \overline{D}$      **d.** $A \overline{B} \overline{C} D$

**6.** To produce a 0 output for the binary number MSB 1011 LSB, the inputs to the NAND gate would be:

   **a.** $\overline{A} \overline{B} \overline{C} D$      **b.** $A \overline{B} C D$

   **c.** $A B \overline{C} D$        **d.** $A \overline{B} \overline{C} \overline{D}$

4-Bit register                                               **Figure 9-13**

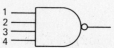

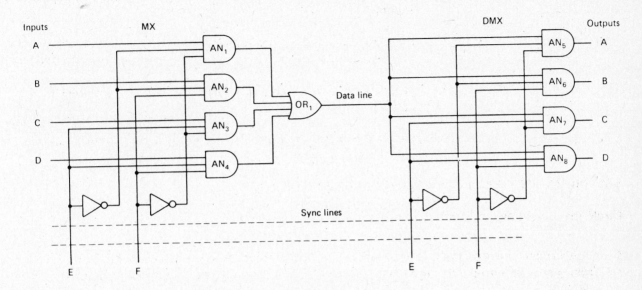

**Figure 9-14**

Questions 7, 8, 9, and 10 refer to Figure 9-14.

7. To select input/output $C$, the condition of $E$ and $F$ inputs would have to be, respectively:

   a. 1 1        b. 1 0

   c. 0 1        d. 0 0

8. The gates that would turn on for question 7 are:

   a. $AN_1$, $OR_1$, and $AN_5$

   b. $AN_2$, $OR_1$, and $AN_6$

   c. $AN_3$, $OR_1$, and $AN_7$

   d. $AN_4$, $OR_1$, and $AN_8$

9. Input $E = 1$ and input $F = 1$, a 1 appears at the output of $AN_4$ and $OR_1$, but none of the outputs contain a 1. The probable cause of trouble is:

   a. $AN_5$        b. $OR_1$

   c. $AN_4$        d. $AN_8$

10. A 1 appears at input $B$ and at output $D$ (all other inputs and outputs are 0). The MX/DMX system appears:

   a. to be okay

   b. to have an open circuit

   c. a short circuit at the inputs

   d. out of synchronization

**ANSWERS TO EXPERIMENTS AND QUIZZES FOR UNIT 9**

*Experiment 1.*

        (1) decimal, binary, or BCD    (2) 0, 1, 1, 0    (3) high, low

*Experiment 2.*

        (1) number    (2) $D$, $\overline{C}$, $B$, $A$

*Experiment 3.*

(1) high     (2) low     (3) 9, low

*Experiment 4.*

(1) 0     (2) 1     (3) odd     (4) input, data     (5) resend

*Experiment 5.*

(1) 8, 1     (2) 1, 8     (3) sequence     (4) sync     (5) proper or correct, 0

*True-False:*

(1) F     (2) F     (3) T     (4) T     (5) T     (6) T     (7) T     (8) F     (9) T
(10) T

*Multiple-Choice:*

(1) d     (2) c     (3) c     (4) a     (5) a     (6) c     (7) b     (8) c     (9) d
(10) d

# Unit 10

# Optoelectronic Indicators and Displays

## SECTION 10-1
### IDENTIFICATION, THEORY, AND OPERATION

### 10-1a  INTRODUCTION

After data have been processed by a digital system, the information must be presented to the outside human world. In Unit 9 it was shown how decoders converted from the binary number system to other number systems, such as decimal. This unit describes the construction of various indicator and display devices and show how they operate.

Light indicators and displays come under that branch of electronics called optoelectronics, where *opto* or *optics* refers to the science of vision and the properties of light. Two fundamental actions of optoelectronics are light detecting or sensing: converting light energy into electrical energy and a light-emitting or light source; and converting electrical energy into light energy. Light indicators and displays employ the latter type of action.

### 10-1b  LIGHT-EMITTING DIODE

A light-emitting diode (LED) is a *pn* junction, as shown in Figure 10-1a. The *p* and *n* materials generally used are gallium phosphide (GaP) or gallium arsenide phosphide (GaAsP). A LED operates in the same manner as a regular diode. In the reverse-biased condition, no current flows and the LED is off. In the forward-biased condition, electrons flow from cathode to anode. When the electrons cross the junction barrier, they fall into holes (a lower energy level) and as a result, photons (particles of light) are given off. This action is referred to as *electron/hole combination* and the LED is on and lighted. Actually, little appreciable current flows or light is produced until the forward-biased voltage is equal to or greater than the inherent for-

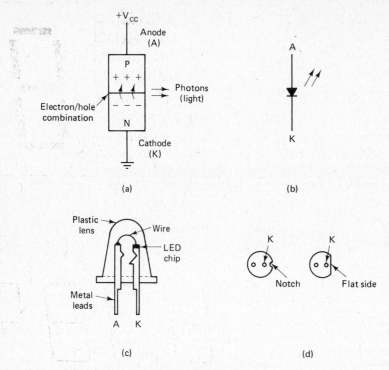

**Figure 10-1** Light-emitting diode: (a) structure and operation; (b) schematic symbol; (c) construction (side view); (d) lead identification (bottom view). (From F. Hughes, *Basic Electronics: Theory and Experimentation,* Prentice-Hall, Englewood Cliffs, N.J., © 1984, Fig. 10-2, p. 213. Reprinted with permission.)

ward voltage drop $(V_F)$ of the LED. This $V_F$ is typically about 2 V. Once the LED is conducting, the $V_F$ is relatively stable, acting somewhat like the reference voltage of a zener diode. The schematic symbol of a LED is shown in Figure 10-1b and the cathode lead identification is shown in Figure 10-1d.

The light emitted by the junction is very small and a glass or epoxy lens is attached above the junction during manufacturing to focus the light, as shown in Figure 10-1c. These lenses may be clear or diffused to give a more visible light source. LEDs come in various sizes and have clear and colored lenses of red, yellow, orange, green, or blue.

A LED can easily be destroyed by excessive current and in most applications must be protected by a series current-limiting resistor $(R_s)$, as shown in Figure 10-2. The value of $R_s$ is easily calculated by using Ohm's law. Since $V_F$ is fairly constant, the voltage across $R_s$ $(V_{R_s})$ is the difference between the voltage applied to the circuit $(V_{CC})$ and $V_F$: $V_{R_s} = V_{CC} - V_F$. A safe current is chosen for the LED that will still produce satisfactory light. This current also flows through $R_s$; therefore, the value of $R_s$ can be found by dividing the current into $V_{R_s}$: $R_s = V_{R_s}/I_F$. This current is the forward-biased current when the LED is on and is indicated as $I_F$.

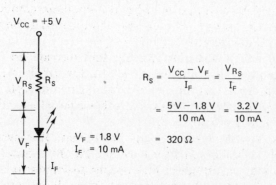

$$R_s = \frac{V_{CC} - V_F}{I_F} = \frac{V_{R_S}}{I_F}$$

$$= \frac{5\,V - 1.8\,V}{10\,mA} = \frac{3.2\,V}{10\,mA}$$

$$= 320\,\Omega$$

$V_F = 1.8\,V$
$I_F = 10\,mA$

**Figure 10-2** Calculation of current-limiting resistor. (From F. Hughes, *Basic Electronics: Theory and Experimentation,* Prentice-Hall, Englewood Cliffs, N.J., © 1984, Fig. 10-3, p. 214. Reprinted with permission.)

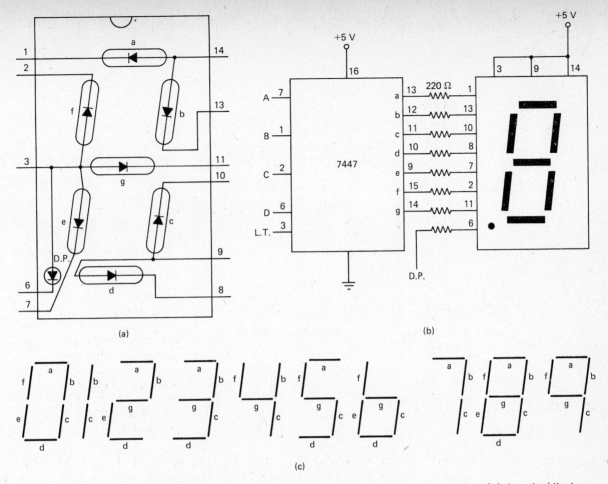

**Figure 10-3** Seven-segment LED display: (a) schematic drawing; (b) decoder/display circuit; (c) segments displaying decimal numbers.

## 10-1c  SEVEN-SEGMENT LED DISPLAY

Seven LEDs are fabricated into a DIP IC package as shown in Figure 10-3a to form what is called a seven-segment LED display. The segments are lettered a through g and selection of various segments will produce the decimal numbers 0 through 9. Proper selection of the segments can also produce the letters A, b, C, d, E, F, H, I, J, L, P, and U. The unit may contain other LEDs to form decimal points and the mathematical operation symbols + and −.

These displays consist of two types, the common anode and the common cathode. With the common-anode type, all anodes are connected to +5 V. The package may accomplish this with the use of one pin or several pins. In the figure, pins 3, 9, and 14 are used with this specific display. External current-limiting resistors are needed with some displays, but others may have them built in. When the various cathodes are connected to ground or 0, the LEDs will light. With the common-cathode type of display, the cathodes are connected to ground and highs or 1s on the anodes of the LEDs will light the segments.

Two styles of LED seven-segment displays are in use. In the older version, each segment was made up of a few LEDs. When a segment was activated, all the LEDs would turn on. A newer style of LED seven-segment display uses solid diffused reflective bars which are connected with light pipes to individual LEDs.

Decoder/drivers are needed to operate the seven-segment LED display properly, as shown in Figure 10-3b. The BCD input is converted to active-low outputs which select the proper segments to be turned on. The input L.T. (pin 3) when brought low will cause all seven segments to turn on, so as to test the decoder and display. The various segments needed to be turned on to represent the decimal numbers 0 through 9 are shown in Figure 10-3c. The segments are generally red, green, orange, or yellow and range in character size from 0.11 to 1.0 inch.

## 10-1d  MULTIDIGIT DISPLAYS

Multidigit displays consist of two or more seven-segment displays contained in a single package or module. Seven inputs select the segments of each digit. Other inputs are used to select each digit. A decimal point is usually included. As each digit is selected, the corresponding data for the digit are sent to the seven segments. An LED concept of this is shown in Figure 10-4a.

**Figure 10-4**  Multidigit multiplexed display: (a) LED concept; (b) single-unit multidigit display; (c) circuit application block diagram.

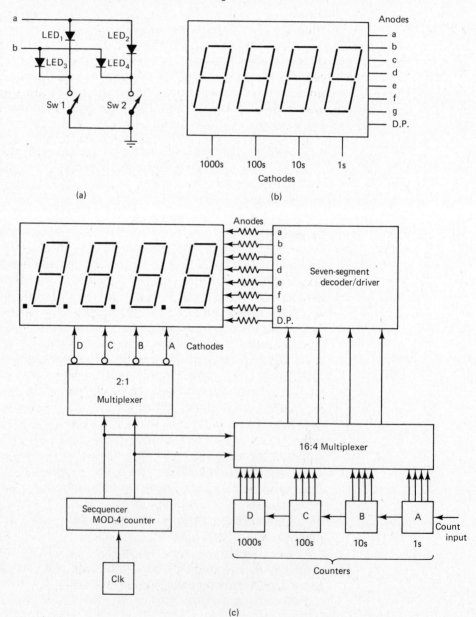

(a)

(b)

(c)

When switch 1 is closed, $LED_1$ and $LED_3$ can turn on, depending on the data at inputs $a$ and $b$. If input $a$ is at 1, $LED_1$ turns on, and if input $b$ is at 1, $LED_3$ turns on. Inputs $a$ and $b$ also connect to $LED_2$ and $LED_4$, respectively, but they will not turn on because switch 2 is open. When Sw 1 is open and Sw 2 is closed, $LED_2$ and $LED_4$ can turn on while $LED_1$ and $LED_3$ will not turn on.

A four-digit display unit is shown in Figure 10-4b. There are seven anode inputs to select the seven segments and four cathode inputs to select each digit. The multidigit display requires a multiplex operation, where the data to the seven anode segments are synchronized with the proper common cathode selected.

A block diagram of a multidigit display system is shown in Figure 10-4c. The outputs of the counters are multiplexed to a single seven-segment decoder/driver which is connected to the seven-segment anode inputs of the display. The output of a sequencer, perhaps using a MOD-4 counter, goes to a multiplexer, which selects each digit and also connects to the counter multiplexer. A system clock (Clk) is used to synchronize the entire circuit. When counter $A$ is selected, cathode line $A$ is also selected on the first digit of the display. The data from counter $A$ will appear on the first digit of the display. When counter $B$ is selected, cathode line $B$ is selected and the data from counter $B$ will appear on the second digit of the display. The operation continues the same way for counters $C$ and $D$ and starts over. The clock frequency is such that the multiplexing rate of the sequencer makes each digit appear to be on continuously because of the slow response of the eye. The result is that each digit display shows the contents of its appropriate counter. A multidigit system such as this reduces power consumption and cost when many-digit-number displays are needed. Some displays of this type may contain all the necessary multiplexing circuits in a single unit.

## 10-1e ALPHANUMERIC DISPLAYS

Alphanumeric displays are able to produce numbers, letters of the alphabet, punctuation marks, and special characters. The LED $5 \times 7$ dot matrix display shown in Figure 10-5a contains 35 LEDs arranged in five columns and seven rows with an extra LED to represent a decimal point (D.P.). In this case, the row inputs are connected to the cathodes of the LEDs and the column inputs are connected to the anodes. To produce a character, the display is scanned by bringing each column high while the various rows are brought low by the data coming from a memory unit. For example, to produce the letter "A" as shown in Figure 10-5b, when column 1 is high (columns 2 through 5 are low), row 1 is low, but rows 2 through 7 are high. When column 2 is high (columns 1 and 3 through 5 are low), rows 1 and 4 are low and rows 2, 3, 5, 6, and 7 are high. The rows remain in the same conditions when columns 3 and 4 each go high. When column 5 goes high, the data to the rows are the same as for column 1. If the columns are scanned, say, 100 times a second as the data are presented to the rows, the LEDs appear to be lighted continuously and the character is produced.

Multisegment alphanumeric displays similar to the seven-segment display are also available, as shown in Figure 10-5c. These 16-segment displays require special decoders or memory devices.

Alphanumeric displays are also found in multidigit units and usually require a multiplexing arrangement. However, the $5 \times 7$ dot matrix display is

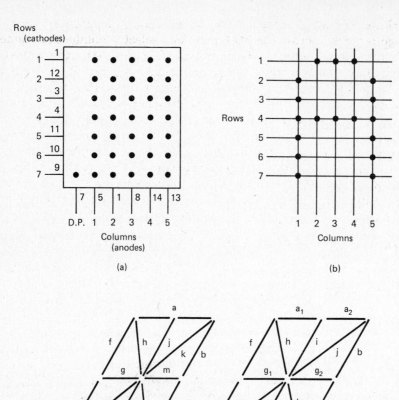

**Figure 10-5** Alphanumeric displays: (a) typical MAN-2 LED 5 × 7 matrix display; (b) example of LED 5 × 7 matrix with the letter "A"; (c) multisegment alphanumeric displays. [(c) After F. Hughes, *Illustrated Guidebook to Electronic Devices and Circuits,* Prentice-Hall, Englewood Cliffs, N.J., © 1983, Fig. 3-19, p. 114. Reprinted with permission.]

more expensive and requires complex control circuits; therefore, it is used only on highly sophisticated systems requiring full alphanumeric capability.

## 10-1f  LIQUID CRYSTAL DISPLAY

A liquid crystal display (LCD or LXD) is different from other displays because it controls light rather than generating light. Its appearance is also different, usually being dark characters against a light or silvery background. However, with some units the characters are light with a dark background.

An exploded view of the construction of a typical LCD display is shown in Figure 10-6a. Because light waves occur in a random fashion, a front polarizing filter is used to allow only light waves oriented in a particular plane, say vertically, to pass through. Next is a plate of glass with transparent conductive segments. Each segment has an electrical edge connector so that various segments can be selected. A nematic fluid (liquid crystal) is placed between the conductive segments plate and a transparent metallized glass back plate. This back plate also has an electrical connector. The last element is the rear polarizing filter that is 90° out of phase, or horizontally oriented with respect to the front filter. The entire unit is sealed during manufacturing.

The liquid crystal material is a semisolid substance and its molecules can be rotated or twisted when subjected to an electrical field. The electrical field placed across the liquid crystal material must be ac or pulsating dc since normal dc voltage will destroy the molecular action in a very short time. This ac voltage is placed across the back plate and selected conductive segments.

When none of the segments are activated, light passes through the front polarizing filter and the transparent conductive segments. The light is then twisted 90° by the liquid crystal, which allows it to pass through the metallized back plate and rear polarizing filter. In many displays the light is then reflected back through the unit and the face of the display will appear blank. When some of the segments are activated, the liquid crystal beneath the segments rotates 90° and the light that passes through these areas is not twisted. When this light reaches the rear polarizing filter, it is absorbed and not allowed to pass through. The result is that the segments on the face of the display will appear dark. There are several methods of LCD operation, but this one is referred to as the *field-effect reflective mode* type.

**Figure 10-6** Liquid crystal display: (a) exploded view of LCD unit; (b) typical LCD display operation.

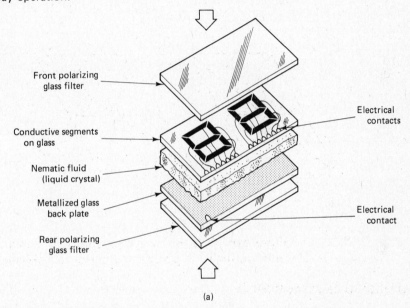

(a)

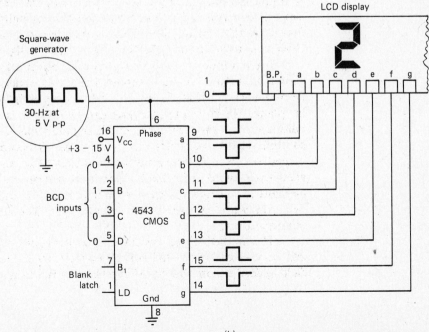

(b)

A basic LCD display operation is shown in Figure 10-6b. Since TTL circuits produce a 0.7-V dc offset voltage at the outputs when conducting, they are not suitable to use with LCD displays. The 4543 IC is a CMOS BCD-to-seven-segment decoder/latch/driver and does not have a dc offset voltage at the outputs. A 30-Hz square-wave pulse train is applied to the back plane (B.P.) of the display and also the phase input (pin 6) of the IC. The number shown at the BCD inputs represent a binary 2. The output stage of the 4543 is an exclusive-OR function resembling a complement circuit, and when the square-wave pulse goes high, the properly selected segments to produce the decimal number 2 go low. This results in a voltage difference between the selected segments and the back plate, causing the liquid crystal to twist under these segments. The unselected segments go high at this time and there is no voltage difference between them and the back plate and therefore no twisting of the liquid crystal. This type of switching action of the output of the 4543 provides clear viewing of the segments without overlapping, where adjacent unselected segments may partially turn on due to capacitive coupling between segments. A high on pin 7 of the 4543 will cause the display to go blank and a low on pin 1 will latch or store the data at the BCD inputs.

Light source displays such as LEDs tend to "wash out" and are difficult to see with an increase in ambient light, whereas the LCD display improves. The LCD display also has lower power operating requirements than the other types of displays.

## 10-1g   OTHER TYPES OF DISPLAYS

### 10-1g.1   Gas Discharge Display

The gas discharge display operates on the principle of the cold-cathode neon glow indicator and is often referred to as a *Nixie tube* (a trademark of the Burroughs Corporation). In one type of numerical display, the cathodes in the shape of numbers 0 through 9 are placed one behind another as shown in Figure 10-7b. A fine-mesh grid type of anode is placed around the cathodes and the entire assembly is in a vacuum tube containing neon gas. Decimal points may also be included in the display. Figure 10-7a shows a front view of the tube displaying the number 3. Figure 10-7c shows the schematic diagram of the gas discharge tube. Notice the current-limiting resistor in the anode circuit to prevent the fine-wire elements from burning out. Approximately 200 V dc is required by the anode, which calls for a special power supply. When one of the numerical cathodes is selected by grounding, electrons flow from cathode to anode. As a result, a typically yellow-orange glow from the gas discharge appears near the surface of the cathode, thus displaying the desired number. Special IC decoder/drivers or transistors are used to energize the cathodes.

Gas discharge displays also come in the form of segment displays and multidigit panels.

### 10-1g.2   Vacuum Fluorescent Display

Fluorescent materials give off a glow (typically green or blue-greenish in color) when struck or bombarded by electrons. The fluorescent material is formed into seven-segment anodes on insulated supporting structures. Filament wires acting as cathodes are placed in front of the anodes and a fine-wire grid may be placed between them. The entire unit is encased in an evacuated glass container. The cathode wires are heated with a low voltage

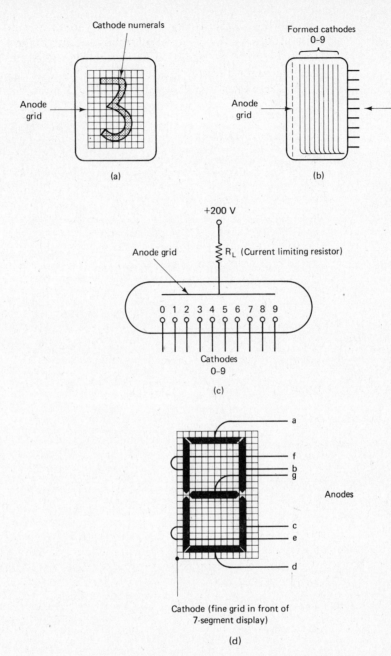

**Figure 10-7** Other types of displays: (a) front view; (b) side view; (c) schematic diagram of gas discharge vacuum tube; (d) vacuum fluorescent display. [From F. Hughes, *Illustrated Guidebook to Electronic Devices and Circuits,* Prentice-Hall, Englewood Cliffs, N.J., © 1983: (a)–(c), Fig. 3-24, p. 117; (d), Fig. 3-20, p. 114. Reprinted with permission.]

and develop a space charge of electrons about them. A higher positive voltage of from 15 to 30 V dc is placed on the grid and selected anode segments. Electrons flow from the filament wires through the grid to the anodes, causing them to glow and show the desired number or character. Electrons are not attracted to unselected anodes and these segments do not glow.

Figure 10-7d shows the front view of a single-digit fluorescent display. These displays are manufactured in single-digit and multidigit units, many of which are used as displays for pocket calculators.

### 10-1g.3  Incandescent Display

Incandescent displays operate, in principle, similar to the ordinary light bulb. Very fine wires in the form of filaments are shaped to characters or seven-segments. These filaments are supported by insulating structures inside a vacuum glass container. When electric current is passed through the selected wires they become hot and glow, thus producing a very bright image of the desired character. These displays have very good viewability in dim or bright light.

## 10-1h  COMPARISON OF DISPLAYS

The following table shows a general comparison of the different displays in relation to *viewability* (the ability to see the display in ambient light conditions), power requirements, *response time* (the time it takes to change from one character to another), and *reliability* (the useful life of the device and resistance to shock).

| Type of Display | Viewability | | Voltage Required | Current Required | Response Time | Reliability |
|---|---|---|---|---|---|---|
| | Dim Light | Bright Light | | | | |
| LED | Good | Poor | 2–4 V dc | 10–20 mA | Fastest | Best |
| LCD | Poor | Best | 1.5–9 V ac | 1–20 $\mu$A | Fair | Good |
| Gas discharge | Good | Fair | 200 V dc | 50 mA | Good | Good |
| Vacuum fluorescent | Good | Fair | 1.5 V ac<br>V dc<br>35 V dc | 10 mA | Good | Good |
| Incandescent | Good | Good | 3–5 V dc<br>V ac | 50–200 mA | Fair | Fair |

## SECTION 10-2
## EXERCISES I

Perform all the exercises in this section before beginning the next section.

1.  Draw the pn structure for a LED and list all parts.

2.  Draw the schematic symbol for a LED, indicating the cathode and anode.

3. Referring to Figure 10-2, find the value of $R_s$ for the following values of $V_{CC}$, $V_F$, and $I_F$.

a. $V_{CC} = +5$ V
   $V_F = +1.8$ V
   $I_F = 15$ mA

   $R_s = $ _____ $\Omega$

b. $V_{CC} = +9$ V
   $V_F = +2$ V
   $I_F = 10$ mA

   $R_s = $ _____ $\Omega$

c. $V_{CC} = +12$ V
   $V_F = +2.1$ V
   $I_F = 20$ mA

   $R_s = $ _____ $\Omega$

d. $V_{CC} = +15$ V
   $V_F = +2$ V
   $I_F = 12$ mA

   $R_s = $ _____ $\Omega$

4. Draw a seven-segment display and label each segment using the letters *a* through *g*.

5. Referring to Exercise 4, list the designated segments that will be activated for the following numbers.

| Decimal Number | Activated Segments |
| --- | --- |
| 0 | |
| 1 | |
| 2 | |
| 3 | |
| 4 | |
| 5 | |
| 6 | |
| 7 | |
| 8 | |
| 9 | |

6. Draw a diagram for a single unit multi-digit display and indicate anode and cathode inputs.

7. Referring to Figure 10-5b, draw a LED 5 × 7 matrix displaying the letter "E".

8. Draw a multisegment alphanumeric display.

9. Draw the exploded view of an LCD display (see Figure 10-6a), and label all parts.

10. Referring to Figure 10-6b, show the level of the seven-segment output of the 4543 decoder/driver IC for the following BDC numbers when the generator pulse is high.

| BCD Number at Input | Seven-Segment Outputs | | | | | | | Decimal Number |
|---|---|---|---|---|---|---|---|---|
| | a | b | c | d | e | f | g | |
| Example:  0 0 0 1 | 1 | 0 | 0 | 1 | 1 | 1 | 1 | 1 |
| 0 0 1 1 | – | – | – | – | – | – | – | |
| 0 1 1 0 | – | – | – | – | – | – | – | |
| 0 1 1 1 | – | – | – | – | – | – | – | |
| 1 0 0 1 | – | – | – | – | – | – | – | |

## SECTION 10-3
## DEFINITION EXERCISES

Give a brief description of each of the following terms.

1. Optoelectronics _____

_____

_____

2. Light detecting _____

_____

_____

3. Light emitting _____

_____

_____

4. LED _____

_____

_____

5. Photons _____

_____

_____

6. Electron/hole combination _____

_____

_____

7. $V_F$ _____

_____

_____

8. Seven-segment display _____

_____

_____

9. Common-anode display _____

_____

_____

10. Common-cathode display _____

_____

_____

11. Multidigit display _____

_____

_____

12. Alphanumeric display _____

_____

_____

13. LCD _____

_____

_____

14. LXD _____

_____

_____

15. Offset voltage _____

_____

_____

16. Gas discharge display _____

_____

_____

17. Nixie tube display _____

_____

_____

18. Vacuum fluorescent display _____

_____

_____

19. Incandescent display _____

_____

_____

20. Viewability _____

_____

_____

21. Response time _____

_____

_____

## SECTION 10-4
## EXPERIMENTS

### EXPERIMENT 1.   BINARY COUNTER WITH SEVEN-SEGMENT DISPLAY

*Objective:*

To show how a display is used to indicate the contents of a counter.

*Introduction:*

A 7493 binary up-counter IC is wired as a MOD-10 or decade counter. The $Q$ outputs are connected to a 7447 BCD-to-seven-segment decoder/driver, which in turn drives a LED seven-segment display.

*Materials Needed:*

*1*   Digital logic trainer *or*
*1*   7493 binary up-counter DIP IC
*1*   7447 BCD-to-seven-segment decoder/driver DIP IC
*1*   MAN-1 seven-segment display or equivalent
*7*   220-$\Omega$ resistors at 0.5 W
*4*   LED indicators
     Several hookup wires or leads

*Procedure:*

1.   Construct the circuit shown in Figure 10-8a.
2.   Activate Sw $A$ until the display shows 0. The counter is now at zero. Record the conditions of the BCD LEDs and the seven segments of the display in binary form on the conversion chart shown in Figure 10-8b.
3.   Activate Sw $A$ for each count and record the new conditions of BCD LEDs and segments of the display in the conversion chart for numbers 1 through 9.

**Figure 10-8**   Binary counter with display: (a) circuit diagram; (b) conversion table.

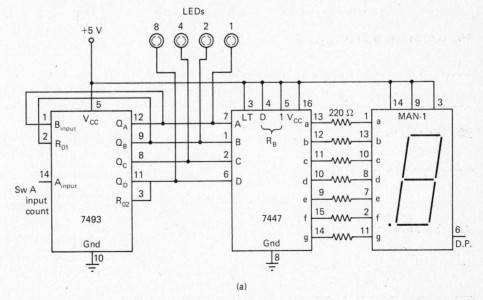

(a)

*(continued)*

**Figure 10-8 (cont.)**

| BCD inputs | | | | Decimal output | 7-segment | | | | | | |
|---|---|---|---|---|---|---|---|---|---|---|---|
| D | C | B | A | | a | b | c | d | e | f | g |
| | | | | 0 | | | | | | | |
| | | | | 1 | | | | | | | |
| | | | | 2 | | | | | | | |
| | | | | 3 | | | | | | | |
| | | | | 4 | | | | | | | |
| | | | | 5 | | | | | | | |
| | | | | 6 | | | | | | | |
| | | | | 7 | | | | | | | |
| | | | | 8 | | | | | | | |
| | | | | 9 | | | | | | | |

(b)

*Fill-In Questions:*

1. When the BCD number is 0101, the lighted segments of the display are

   _____.

2. When the BCD number is 1000, the lighted segments of the display are

   _____.

3. When segments *a*, *b*, *c*, *d*, and *g* are lighted, the decimal number _____

   _____is displayed.

4. When segments *a*, *c*, *d*, *e*, *f*, and *g* are lighted, the decimal number _____

   _____is displayed.

**SECTION 10-5
INSTANT REVIEW**

- A LED operates in the same manner as a regular diode, but will be lighted in the forward-biased condition. Its normal $V_F$ is about 2 V, but it usually requires a current-limiting resistor in series with it.
- A LED display has good viewability in dim light but poor viewability in bright light, has the fastest response time, and is the best of the displays for reliability.
- The LCD display requires ac or pulsating dc voltage across its back plate and segments to operate, and CMOS decoder/drivers are used since they have no dc offset voltages. The LCD display viewability is poor in dim light, the best of all the other displays in bright light, has a fair response time and good reliability, and uses the least amount of power.
- The gas discharge display requires a high dc operating voltage. Its viewability is good in dim light but poor in bright light, it has good response time, and it has good reliability.
- The vacuum fluorescent display requires a filament voltage and a higher

dc anode voltage to operate. Its viewability is good in dim light but fair in bright light, it has good response time, and it has good reliability.

- The incandescent display uses the most power, but is the brightest display for all light conditions. It has fair response time and reliability.


**SECTION 10-6**
**EXERCISES II**

Perform all the exercises in this section before beginning the next section.

1. List the components of a multidigit display system used with counters and describe briefly the function of each component (refer to Figure 10-4b).

   *Component*                     *Description*

   1.

   2.

   3.

   4.

   5.

   6.

2. List the type of display based on its power requirements from lowest to highest.

   | Power Requirement | Type of Display |
   |---|---|
   | Lowest | |
   | Low | |
   | Medium | |
   | High | |
   | Highest | |

3. Match the type of display in column A with its proper description in column B.

   | *Column A* | | *Column B* | |
   |---|---|---|---|
   | ____ 1. | LED | a. | Requires the highest operating voltage. |
   | ____ 2. | LCD | b. | Is the brightest for all light conditions. |
   | ____ 3. | Gas discharge | c. | Uses the least amount of power. |
   | ____ 4. | Vacuum fluorescent | d. | Has the fastest response time. |
   | ____ 5. | Incandescent | e. | Display is green or blue-green in color when operating. |

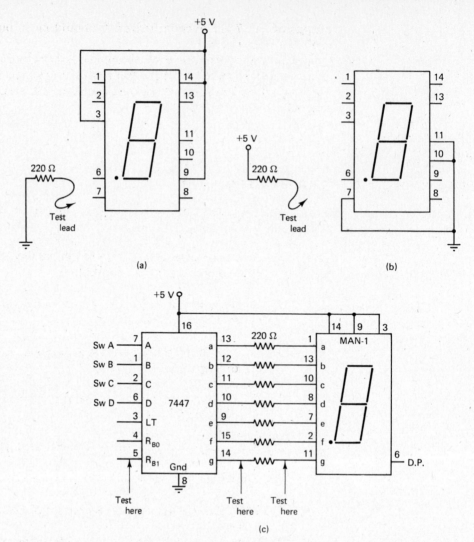

**Figure 10-9**  Testing displays: (a) common-anode seven-segment; (b) common-cathode seven-segment; (c) decoder/driver display system.

## SECTION 10-7
## TROUBLESHOOTING APPLICATION:
## TESTING DISPLAY SYSTEMS

It may be necessary to test a seven-segment display before placing it in a circuit. For a common-anode display, connect all anodes to the +5-V supply as shown in Figure 10-9a. Then using a 220-$\Omega$ resistor with one end connected to ground, touch each cathode pin to light the various segments.

For a common-cathode display, connect all cathodes to ground as shown in Figure 10-9b. Then using a 220-$\Omega$ resistor with one end connected to the +5-V supply, touch each anode pin to light the various segments.

An "in-circuit" test for a display system means checking the inputs to the display unit as shown in Figure 10-9c. If the light test (pin 3) of the 7447 IC is brought low, all seven segments of the display should light, showing the number 8. For this particular display, all inputs should be low; similarly, all outputs of the 7447 decoder/driver should be low. If one segment is off, check to see the data at their corresponding input to the display. If this input is 0, the display is faulty. If this input is 1, check the corresponding output of the 7447 decoder/driver. It should be 0, but if it is 1, the 7447 is probably faulty.

If all segments light with the previous test, but the wrong number appears on the display during normal operation, check the BCD inputs to the 7447 decoder/driver IC to make sure that the correct number is being entered.

## SECTION 10-8
## SELF-CHECKING QUIZZES

### 10-8a  DIGITAL DISPLAYS: TRUE-FALSE QUIZ

Place a T for true or an F for false to the left of each statement.

_____  1.  A resistor is usually connected in parallel with a LED to limit the current through it to a safe value.

_____  2.  A gas discharge display has an orange-yellow glow when it is operating.

_____  3.  The vacuum fluorescent display consumes the least current of all the displays.

_____  4.  A common cathode seven-segment display requires 1s to turn on its segments.

_____  5.  A multidigit display requires multiplexing for proper operation.

_____  6.  Alphanumeric displays include dot-matrix or segment types.

_____  7.  LCD displays have faster response time than LED displays.

_____  8.  Incandescent displays require high operating voltage.

_____  9.  LCD displays must operate with ac or pulsating dc 180° out of phase from the back plate to the desired segments.

_____  10.  LED displays are best seen in bright sunlight.

### 10-8b  DIGITAL DISPLAYS: MULTIPLE-CHOICE QUIZ

Circle the correct answer for each question.

1.  The flat side or notch on the case of a LED indicates the:

a.  voltage rating  b.  cathode

c.  anode  d.  manufacturer

2.  A seven-segment display can be made of:

a.  LED

b.  LCD

c.  gas discharge

d.  vacuum fluorescent

e.  incandescent

f.  all of the above

3. The type of display that gives off a green or blue-greenish light is the:

    a. LED

    b. LCD

    c. gas discharge

    d. vacuum fluorescent

4. A common feature to all displays is the:

    a. emission of light

    b. control of light

    c. display of information

    d. none of the above

5. A display that can be used for alphanumeric information is the:

    a. LED 16-segment display

    b. Nixie tube decimal display

    c. 5 × 7 LED dot-matrix display

    d. LCD seven-segment display

    e. both answers a and c

6. If the number 5 is shown on a seven-segment display, the segments activated are:

    a. $a$, $b$, $g$, $e$, and $d$

    b. $f$, $g$, $e$, $c$, and $d$

    c. $a$, $f$, $g$, $c$, and $d$

    d. $a$, $b$, $c$, $d$, and $g$

7. In a LED circuit, $V_{CC} = +6$ V, $V_F = 1.8$ V, and $I_F = 12$ mA. The best choice for the value of $R_s$ is:

    a. 100 $\Omega$          b. 220 $\Omega$

    c. 330 $\Omega$          d. 1k $\Omega$

8. The best display to use in bright sunlight is the:

    a. LED display

    b. LCD display

    c. vacuum fluorescent display

    d. gas discharge display

9. The display that requires only a filament voltage is the:

    a. LED

    b. LCD

    c. vacuum fluorescent

    d. incandescent

    e. gas discharge

10. In a counter display, if one segment remains off in a seven-segment display for all input numbers, the first place to test would be the:

    a. BCD inputs to the counter

    b. $Q$ outputs of the counter

    c. inputs to the decoder/driver

    d. inputs to the seven-segment display

## ANSWERS TO EXPERIMENTS AND QUIZZES FOR UNIT 10

*Experiment 1:*

(1) $a$, $f$, $g$, $c$, $d$    (2) $a$, $b$, $c$, $d$, $e$, $f$, $g$    (3) 3    (4) 6

*True-False:*

(1) F    (2) T    (3) F    (4) T    (5) T    (6) T    (7) F    (8) F    (9) T
(10) F

*Multiple-Choice:*

(1) b    (2) f    (3) d    (4) c    (5) e    (6) c    (7) c    (8) b    (9) d
(10) d

# Unit 11

---

# Memory Devices

## SECTION 11-1
## IDENTIFICATION, THEORY, AND OPERATION

### 11-1a  INTRODUCTION

In Unit 5 it was shown how a flip-flop could store a 1 or 0 bit, making it a basic memory device. A group of flip-flops can be combined into a register, as presented in Unit 6, and is capable of storing a group of bits called a *data word*. Semiconductor memories use flip-flops as memory cells arranged in a memory array. Depending on the organization of a memory device, it is capable of storing many words. Memory devices also require additional circuits such as decoders, control gates, sense amplifiers and drivers, so the proper memory locations can be selected or accessed where data can be stored (write-in) and retrieved (read-out).

This unit will concentrate mainly on semiconductor memories in IC form, but other types of memory devices will be mentioned briefly to show the other types in use.

### 11-1b  THE RANDOM ACCESS MEMORY

A *random access memory* (RAM) is a memory device where any one or group of memory cells can be selected (accessed) directly. This type of memory can have data written into and read out of it; therefore, it is also referred to as a *read/write memory*. The specific group of cells in memory is called a *memory location*. Figure 11-1a shows a block diagram of the 7489 RAM DIP IC. This memory device is organized into 16 words of 4 bits each, making a total of 64 bits or memory cells in the memory array. Only word 0 and word 15 are shown to simplify the drawing. In a sense, the memory array contains 16 binary registers, each with a word 4 bits long.

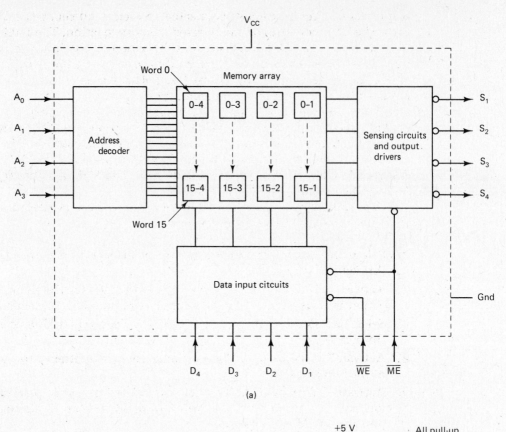

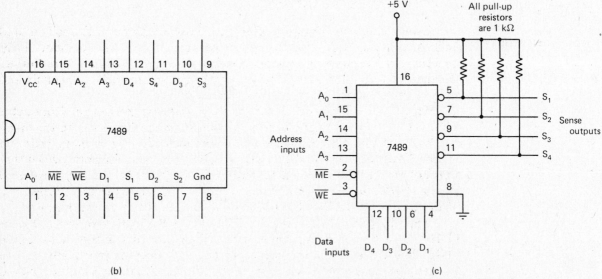

**Figure 11-1**  7489 16 × 4-bit RAM DIP IC: (a) general block diagram; (b) pin identification; (c) circuit arrangement with pull-up resistors.

The address inputs $A_0$ through $A_3$ are connected to the address decoder which is capable of accessing any specific word at random. For example, if all inputs are 0, word 0 is selected, and if all inputs are 1, word 15 is selected. The address inputs are usually connected to an external temporary storage register called an *address register*.

The data inputs $D_1$ through $D_4$ are connected to each word cell, respectively; $D_1$ is connected to cell 0-1 through cell 15-1, and so on. The sensing amplifiers or circuits are also connected to each word cell, respectively, and show the data contents of the selected memory location. The bubbles at the outputs of $S_1$ through $S_4$ indicate that the data read out are the complement of the data written into the memory. For example, if the word stored was 1101, the sense outputs would be 0010. This feature is useful in reducing the need for other ICs in certain applications.

Two other inputs control the operation of the memory IC: memory enable (ME) and write enable (WE). When ME is low, the IC is activated and the sense outputs will show the complemented data stored at the memory location selected by the $A_0$ through $A_3$ address inputs. With ME low, when WE goes low it will write in or store the new data at data inputs $D_1$ through $D_4$ into the memory location selected.

The following step-by-step procedure shows how to write in and read out data from any desired memory location.

1. Select the desired memory location by placing 1s and 0s at inputs $A_0$ through $A_3$.
2. Activate ME input low. The old data stored at the selected memory location will appear at the $S_1$ through $S_4$ outputs in complemented form.
3. Select the desired data to be stored by placing 1s and 0s at inputs $D_1$ through $D_4$.
4. Activate WE input low. The new data are now written into the selected memory location.
5. Activate WE input high. The new data stored into the selected memory location can be read out and will now appear at the $S_1$ through $S_4$ outputs in complemented form.

Even though the data stored in a memory location can be read out and used for other applications, the original data remain in that location. This is sometimes referred to as a *nondestructive readout*. In other memory devices the data at a memory location are changed or destroyed on readout and must be restored with a write-in operation.

There are two ways data can be altered or changed in a semiconductor RAM: (1) by the writing in of new data, or (2) when the power to the memory is lost. When the power returns to the memory, the memory cells will assume different states and the data will be garbled. A memory device that does not retain its data when power is removed from the memory is referred to as a *volatile memory*. A *nonvolatile memory*, such as magnetic types, will retain the correct data for a certain length of time after power is removed.

The pin identification for the 7489 64-bit RAM IC is shown in Figure 11-1b and a circuit arrangement with pull-up resistors is shown in Figure 11-1c. The sense amplifiers have open collector outputs (not connected to $V_{CC}$) to permit *wired-OR capability*, which means that the sense outputs of another memory IC can be connected to these lines without causing interference between the two devices. Therefore, a 1-k$\Omega$ pull-up resistor is required between $V_{CC}$ and each $S_1$ through $S_4$ output to be able to pull the output up toward $V_{CC}$ when the output state is 1.

## 11-1c  INCREASING THE WORD LENGTH OF A MEMORY

The word length of a memory can be increased by connecting memory ICs together. Figure 11-2 shows how two 7489 RAM ICs can be connected to produce a 16 $\times$ 8-bit RAM. The address inputs $A_0$ through $A_3$ of each IC

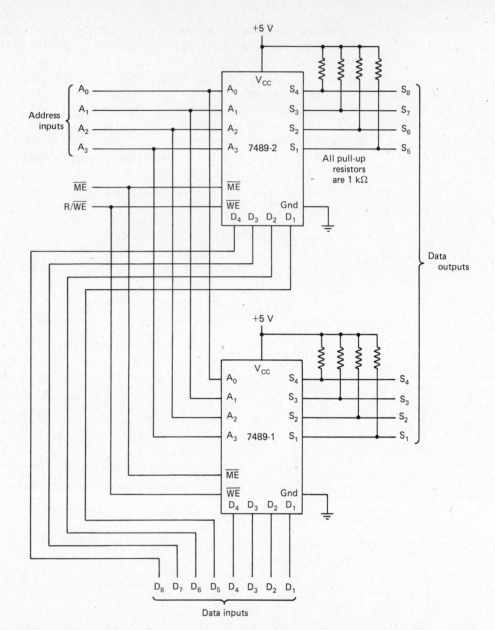

**Figure 11-2**  Increasing the word length to produce a 16 × 8-bit RAM.

are connected in parallel. With this arrangement the same memory location is selected in each IC at the same time. Inputs ME and WE are also connected in parallel on each IC. This allows both ICs to be enabled and written into at the same time. The data lines of each IC are brought out separately to form an 8-bit data input. Similarly, the sense outputs of each IC are brought out separately to form an 8-bit data output. Pull-up resistors are needed with each output. The read/write operation is the same as for a single 7489 IC, except that there are eight data inputs and eight sense outputs, producing a 16 × 8-bit RAM.

## 11-1d  INCREASING THE NUMBER OF WORDS IN A MEMORY

The number of words in a memory can also be increased by connecting memory ICs together as shown in Figure 11-3. The address inputs are connected together as before, but the ME input of IC 7489-1 acts as the fifth address input. As long as this input is 0, only the 7489-1 IC is enabled and the first 16 words of the memory are accessible. The other IC, 7489-2, will not be enabled because of the inverter at its ME input; therefore, none of the memory locations of this IC can be selected. When $A_4$ goes to a 1, IC 7489-1 is disabled and IC 7489-2 is enabled. Even though the first four address inputs are still used, only the memory locations of IC 7489-2 are accessible. The four data lines of each IC are connected together and also the four sense lines of each IC are connected together. Since both ICs have open collector outputs, only four pull-up resistors are used and shared by each IC. This represents wired-OR connection. The result of this arrangement is a $32 \times 4$-bit RAM.

**Figure 11-3**   Increasing the number of words to produce a $32 \times 4$-bit RAM.

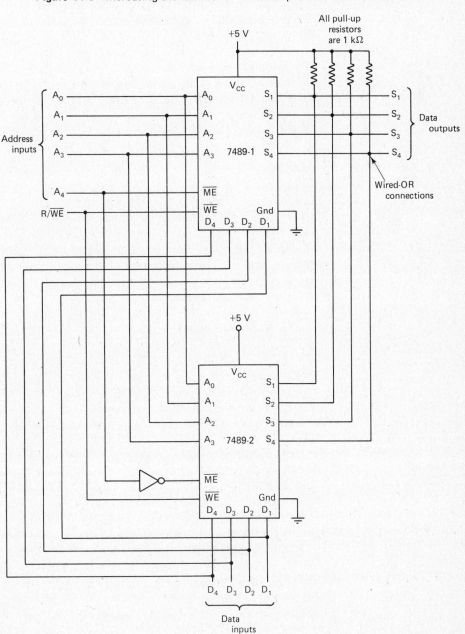

## 11-e  ORGANIZATION OF OTHER MEMORY ICs

There are several other types of organization for memory ICs, some of which are:

| | |
|---|---|
| 256 × 1-bit memory | 2048 × 8-bit memory |
| 256 × 4-bit memory | 4096 × 1-bit memory |
| 1024 × 1-bit memory | 16384 × 1-bit memory |
| 1024 × 4-bit memory | 32768 × 1-bit memory |
| 1024 × 8-bit memory | 65536 × 1-bit memory |

The use of these other memory ICs make it more efficient from the standpoint of construction, operation, and cost when producing a larger memory. For example, to construct a 256 × 8-bit RAM, thirty-two 7489 ICs are needed, whereas only eight 256 × 1-bit RAM ICs can be used, or better still, two 256 × 4-bit ICs can be used.

Generally, it should be remembered that to increase the word length of a memory, the address inputs are wired together in parallel and the data inputs and outputs are wired separately. However, to increase the number of words in a memory, the data inputs and outputs are connected in parallel and the address inputs, although connected in parallel, are arranged so that only one IC is enabled at a time with the ME input.

## 11-1f  THE READ-ONLY MEMORY

The *read-only memory* (ROM) is a memory from which data can be read out repeatedly, but cannot be written into. The ROM is used where the same data and information are used quite often, such as code conversions, reference tables used in mathematical operations, character generation for output devices, and monitor programs used to operate computers. Since the information stored in a ROM does not change when power is removed, it is considered a nonvolatile memory.

### 11-1f.1  Factory-Programmed ROM

When the customer orders a ROM from the manufacturer, the information as to the type of data to be stored is also supplied. During the fabricating process specific memory cells are selected for 1s and 0s. Figure 11-4a shows how one MOSFET has a gate representing a 1 when the select line is activated, and the other MOSFET has no gate representing a 0.

### 11-1f.2  Field-Programmable ROM

The field-programmable ROM, usually called a FROM or PROM, differs from the ROM in that the customer or user is able to program the memory with a special programming device. An example of a bipolar PROM is shown in Figure 11-4b. The transistors that make up the outputs of the memory cells are designed with fuse links in their emitter leads. A new PROM received by the user will have all its on-chip fuse links intact, and the outputs can be all 1s or 0s, depending on the type of memory needed. The user places the PROM into the programming device and applies a sufficiently high voltage to "blow" the desired fuse links. A blown fuse link may indicate a 1

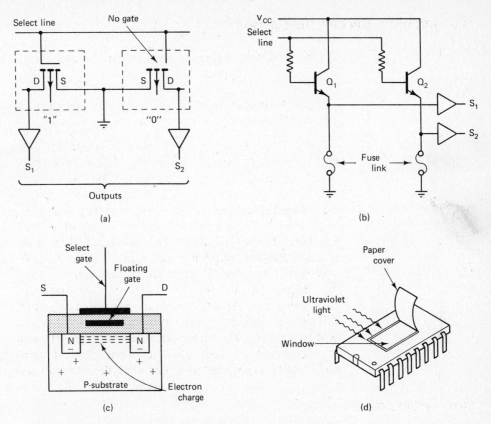

**Figure 11-4**   ROM structures: (a) factory-programmed ROM; (b) field-programmable ROM; (c) erasable PROM; (d) EPROM package with window.

or 0 as datum, which again depends on the type of PROM used. The main disadvantage with this procedure is that if a mistake is made during the programming operation, the entire memory IC must be discarded.

### 11-1f.3   Erasable PROM

The erasable PROM, also called an EPROM, is a ROM that can be programmed by the user for an application that will be used repeatedly for a specific length of time and then erased and reprogrammed for another job. An EPROM is ideal for engineering applications where specific programs are being tested before the final program is approved, and then can be placed in a permanent ROM.

The EPROM consists of metal-oxide semiconductors (MOS) structures, as shown in Figure 11-4c. A floating gate is located in the insulating silicon dioxide layer of the MOSFET. During the programming operation a higher-than-normal voltage is applied to the select gates of the cells that are to indicate a 0 output. The floating gate assumes a charge that for all practical purposes requires a very long time to discharge under normal circumstances. With normal read-out conditions, a lower voltage is placed on the select gate. The MOSFETs with a charged floating gate will turn on, and those without the charge will remain off.

To erase the data stored in the EPROM, the floating gates must be discharged. This is accomplished by allowing ultraviolet light to enter a special quartz window on top of the IC, as shown in Figure 11-4d. The charges on the floating gates are forced back into the substrate of each MOSFET, and the memory can be programmed again with special PROM programming equipment.

### 11-1f.4  Electrically Alterable PROM

The electrically alterable PROM (EAPROM) or electrically erasable PROM (EEPROM) is similar to the normal EPROM, but is programmed and erased electrically. Even though it has a slow write time compared to its read time, it could become the nonvolatile RAM of the future.

## 11-1g  OTHER TYPES OF MEMORIES

The semiconductor 7489 RAM studied in Sections 11-1a to 11-1d is referred to as a *static memory*. The data stored in its memory cells are fixed and can be read out repeatedly unless new data are written in or power to the memory is lost.

There also exist dynamic memories, where data are stored as electrical charges on capacitors. These memories use MOS technology and the small capacitors are fabricated together with the MOSFETs. Since the stored charges decay in a fraction of a second, they have to be *refreshed* (charged up) very often by some method—hence the term *dynamic*, which implies energy in motion. The main advantage of this type of memory is that more memory cells can be packed into a given area of the chip.

Another dynamic memory is the *charged-coupled device* (CCD). This type of memory uses a substrate and a series of small metal gates separated by an insulating oxide layer. A small positive charge injected into the substrate forms a so-called "bucket," representing a 1. With sequential activation of the gates, the charged bucket can be moved from input to output, similar to a shift register.

Magnetic-type memories are also in use with digital circuits. The older magnetic-core type consists of small doughnut-shaped *ferrite cores* smaller in diameter than the lead of a normal pencil. A single ferrite core will hold one bit of data. These cores are grouped in memory arrays with very fine wires running through them. When current is passed through the wires, electromagnetism will magnetize the cores in various directions. When the current is reversed in some wires, the magnetic fields of the cores will switch in the opposite direction and give an indication of the stored data. A second operation is needed to restore the magnetic fields to their original condition and is called a *rewrite operation*. This type of memory is called a *destructive readout*, since the original data are lost during the readout operation and must be rewritten back into memory. The best feature of the magnetic core memory is that it is nonvolatile and will retain its magnetized condition when power is lost.

Magnetic bubble memory is a newer type of memory that uses tiny magnetic domains in a thin film of magnetic material. These domains or bubbles can be produced and moved around by permanent magnets and coils built around the magnetic material.

Magnetic materials also have widespread applications for auxiliary or external memory in the form of magnetic drum, reel-to-reel recording tape, cassettes, normal disks, and floppy disks. All of these devices are required to be moving or rotating in order to store and retrieve the data.

Punched cards and punched paper tape are also forms of external memory.

## 11-1h  OTHER MEMORY TERMINOLOGY

The following terms are usually associated with memory devices.

**Access Time:**  This is the time, in microseconds or nanoseconds, that it takes to write and read data of a memory.

**Address (noun):**  A specific location in memory where data can be stored.

**To Address (verb):**  The process of selecting a desired memory location for writing in or reading out data from a memory.

**Address Register:**  A register that temporarily holds the data needed to select a specific address in the memory.

**Memory Data Register:**  A register that temporarily holds the data to be stored into memory or receives data coming from memory.

**Program (noun):**  A set of instructions and/or data that are stored in a memory that controls the operations of a digital system.

**To Program (verb):**  The procedures for storing a program into a memory.

**Software:**  The written instructions and operating procedures for a digital system. A program written on paper.

**Firmware:**  The instructions and data that are stored in ROM, such as a monitor program for a computer, mathematical tables, and conversion programs.

**Hardware:**  The actual ICs, resistors, capacitors, PC boards, and physical parts of a digital system.

## SECTION 11-2
## EXERCISES I

Perform all the exercises in this section before beginning the next section.

1.  Draw the block diagram for a 7489 RAM IC (refer to Figure 11-1a) and label all parts.

2.  Draw the circuit arrangement for a 7489 RAM IC (refer to Figure 11-1c).

3. When increasing the word length using 7489 RAM ICs, indicate which inputs/outputs remain the same and those that increase in total number.

| Leads | Total Number |
|---|---|
| Address inputs | _____ |
| Data inputs | _____ |
| Sense outputs | _____ |

4. When increasing the number of words using 7489 RAM ICs, indicate which inputs/outputs remain the same and those that increase in total number.

| Leads | Total Number |
|---|---|
| Address inputs | _____ |
| Data inputs | _____ |
| Sense outputs | _____ |

5. Using 7489 ICs, indicate how many ICs must be used to produce the following memories.

| Type of RAM | Number of ICs Needed |
|---|---|
| a. 64 × 4-bit | _____ |
| b. 32 × 8-bit | _____ |
| c. 128 × 4-bit | _____ |
| d. 128 × 8-bit | _____ |
| e. 256 × 8-bit | _____ |

6. Place the following procedures in proper order for writing in and reading out data with a 7489 RAM IC.

| Order of Sequence | Operations |
|---|---|
| 1. _____ | a. Activate ME input low |
| 2. _____ | b. Enter data at data inputs |
| 3. _____ | c. Enter data at address inputs |
| 4. _____ | d. Activate WE input low |
| 5. _____ | e. Activate WE input high |
| 6. _____ | f. Observe sense outputs |

**SECTION 11-3**
**DEFINITION EXERCISES**

Give a brief description of each of the following terms.

1. Memory cell _____

_____

_____

2. Memory array _____

_____

_____

3. Access _____

_____

_____

4. Access time _____

_____

_____

5. Write in _____

_____

_____

6. Read out _____

_____

_____

7. Memory location _____

_____

_____

8. Memory enable input _____

_____

_____

9.  Write enable input _____

_____

_____

10. Address decoder _____

_____

_____

11. Data inputs _____

_____

_____

12. Sense outputs _____

_____

_____

13. Open collector outputs _____

_____

_____

14. Pull-up resistors _____

_____

_____

15. Wired-OR connections _____

_____

_____

16. RAM _____

_____

_____

17. ROM _____

_____

_____

18. PROM _____

_____

_____

19. EPROM _____

_____

_____

20. EEPROM _____

_____

_____

21. Volatile _____

_____

_____

22. Nonvolatile _____

_____

_____

23. To store _____

_____

_____

24. To retrieve _____

_____

_____

25. Static memory _____

_____

_____

26. Dynamic memory _____

_____

_____

27.  Magnetic core memory _____

_____

_____

28.  Magnetic bubble memory _____

_____

_____

29.  Destructive readout _____

_____

_____

30.  Nondestructive readout _____

_____

_____

31.  CCD _____

_____

_____

32.  Auxiliary memory _____

_____

_____

33.  Address (noun) _____

_____

_____

34.  To address (verb) _____

_____

_____

35.  Address register _____

_____

_____

36. Memory data register _____

_____

_____

37. Program (noun) _____

_____

_____

38. To program (verb) _____

_____

_____

39. Software _____

_____

_____

40. Firmware _____

_____

_____

41. Hardware _____

_____

_____

## SECTION 11-4
## EXPERIMENTS

The two experiments in this section involve constructing small digital systems using the 7489 RAM IC. It may be advantageous if Section 11-7 on testing a 7489 RAM is performed initially to gain experience with memory ICs and to be sure that the devices are functioning properly. Since the wiring is more complex and requires resistors, these experiments could be wired "hard" on modular perforated boards. The learner only has to apply a +5 V power supply and perform the experiments. (*Suggestion:* All ICs should be mounted in IC sockets.)

**EXPERIMENT 1.  BINARY-TO-HEXADECIMAL DECODER DISPLAY**

*Objective:*

To show how two 7489 RAM ICs can be used to convert binary numbers to hexadecimal numbers.

*Introduction:*

Two 7489 RAM ICs are connected to produce a 16 × 7-bit word memory. A 7493 binary up-counter is used to represent the binary numbers and to select the various memory locations. Initially, each memory location is stored with the proper data, via the data switches, to indicate the proper hexadecimal number on the seven-segment display. After all 16 numbers are programmed in the memory, the counter is advanced and a comparison is made between the four LED indicators (binary) and the seven-segment display (hexadecimal).

*Materials Needed:*

1   Digital logic trainer *or*

1   7493 binary up-counter DIP IC

2   7489 64-bit RAM DIP ICs

1   MAN-1 seven-segment display or equivalent

7   220-Ω resistors at 0.25 W

7   1-kΩ resistors at 0.25W

4   LED indicators

Several hookup wires or leads

 *or*

1   Complete hard-wired module and +5 V regulated power supply

*Procedure:*

1.  Connect the circuit as shown in Figure 11-5a.
2.  Make sure that Sw *H* ($\overline{\text{WE}}$) is high.
3.  Using Sw *I*, advance the counter until all LED indicators are 0.
4.  Set data switches *A* through *G* as indicated by the first row in the programming chart shown in Figure 11-5b.
5.  Momentarily activate Sw *H* ($\overline{\text{WE}}$) low and then high. The data for the seven-segment display for number 0 have been stored in memory.
6.  Continue the same procedure as in steps 4 and 5 for the remaining numbers in the programming chart.
7.  Once all the numbers have been programmed into the memory, make sure that Sw *H* ($\overline{\text{WE}}$) is high. Now activate Sw *I* and compare the four binary LED indicators with the seven-segment display.
8.  If there are any mistakes, refer to the programming chart and set up the proper data switches. Then activate Sw *H* ($\overline{\text{WE}}$) low and then high to store the proper data.

*Fill-In Questions:*

1.  When the binary number is 1010, the seven-segement display data are

   a =_____, b =_____, c =_____, d =_____

   _____, e =_____, f =_____, and g =_____

   _____, with the hexadecimal number _____ appearing

   on the display.

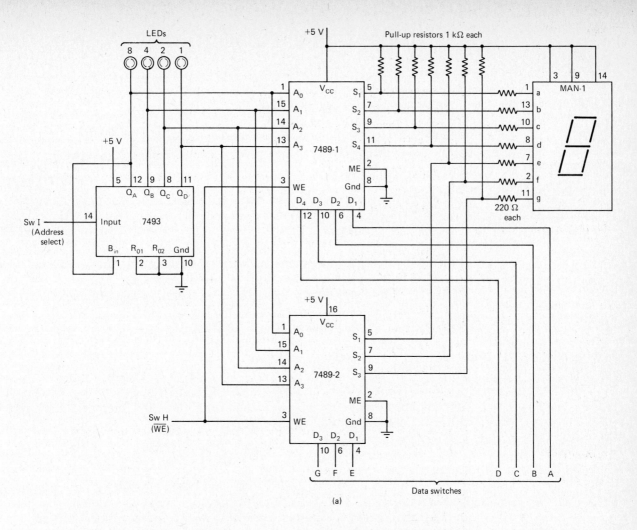

| | | | | 7-segments | | | | | | | |
|---|---|---|---|---|---|---|---|---|---|---|---|
| Binary | | | | a | b | c | d | e | f | g | Hexadecimal |
| 0 | 0 | 0 | 0 | 1 | 1 | 1 | 1 | 1 | 1 | 0 | 0 |
| 0 | 0 | 0 | 1 | 0 | 1 | 1 | 0 | 0 | 0 | 0 | 1 |
| 0 | 0 | 1 | 0 | 1 | 1 | 0 | 1 | 1 | 0 | 1 | 2 |
| 0 | 0 | 1 | 1 | 1 | 1 | 1 | 1 | 0 | 0 | 1 | 3 |
| 0 | 1 | 0 | 0 | 0 | 1 | 1 | 0 | 0 | 1 | 1 | 4 |
| 0 | 1 | 0 | 1 | 1 | 0 | 1 | 1 | 0 | 1 | 1 | 5 |
| 0 | 1 | 1 | 0 | 1 | 0 | 1 | 1 | 1 | 1 | 1 | 6 |
| 0 | 1 | 1 | 1 | 1 | 1 | 1 | 0 | 0 | 0 | 0 | 7 |
| 1 | 0 | 0 | 0 | 1 | 1 | 1 | 1 | 1 | 1 | 1 | 8 |
| 1 | 0 | 0 | 1 | 1 | 1 | 1 | 0 | 0 | 1 | 1 | 9 |
| 1 | 0 | 1 | 0 | 1 | 1 | 1 | 0 | 1 | 1 | 1 | A |
| 1 | 0 | 1 | 1 | 0 | 0 | 1 | 1 | 1 | 1 | 1 | B |
| 1 | 1 | 0 | 0 | 1 | 0 | 0 | 1 | 1 | 1 | 0 | C |
| 1 | 1 | 0 | 1 | 0 | 1 | 1 | 1 | 1 | 0 | 1 | D |
| 1 | 1 | 1 | 0 | 1 | 0 | 0 | 1 | 1 | 1 | 1 | E |
| 1 | 1 | 1 | 1 | 1 | 0 | 0 | 0 | 1 | 1 | 1 | F |

(b)

**Figure 11-5**  Binary-to-hexadecimal decoder using two 7489 RAM ICs: (a) circuit; (b) programming chart.

2. The _____ inputs are activated _____ simultaneously to write data into the memory.

3. The 1-kΩ pull-up resistors are used because the 7480 RAM ICs have _____ _____ outputs.

4. The 220-Ω resistors are used to _____ _____ to the seven-segment display LEDs.

5. The ME inputs are permanently wired low; therefore, the seven-segment display shows the data in the memory location selected by the data at the _____ inputs.

## EXPERIMENT 2.   ALPHANUMERIC 5 X 7 DOT MATRIX DISPLAY

*Objective:*

To show how a memory is programmed to produce alphanumeric characters on a 5 × 7 dot matrix display.

*Introduction:*

The particular character that you wish to display is first planned on the matrix shown in Figure 11-6b. This datum is then entered into the first five memory locations. The 7493 IC is wired as a MOD-6 counter. When the counter input is set to the clock (Clk), the five memory locations are accessed in a sequential manner, sending their inverted data to the seven rows of the 5 × 7 display. At the same time, the outputs of the counter are also sent to a 7442 BCD-to-decimal decoder. These outputs go low in a scanning manner, which are then inverted by the 7404 hex IC and each column of the 5 × 7 display goes high the same time the proper data are sent to the rows. The clock speed is adjusted so that, to the human eye. the character appears stationary.

*Materials Needed:*

1   Digital logic trainer *or*
1   7493 binary up-counter DIP IC
2   7489 64-bit RAM DIP ICs
1   7442 BCD-to-decimal decoder DIP IC
1   7404 hex inverter DIP IC
1   MAN-2 LED 5 × 7 dot matrix display or equivalent
5   220-Ω resistors at 0.25 W
7   1-kΩ resistors at 0.25 W
3   LED indicators
    Several hookup wires or leads
        *or*
1   Complete wired-hard module and +5 V regulated power supply

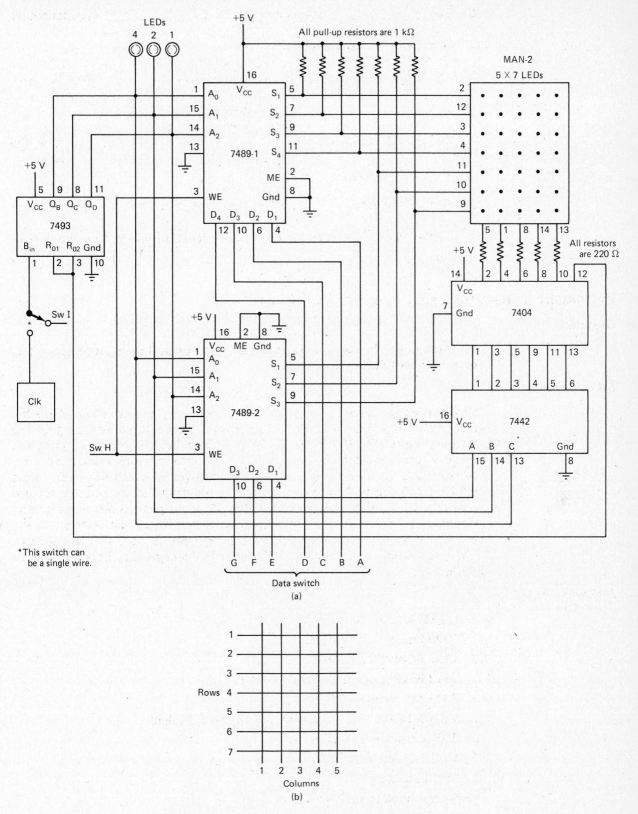

**Figure 11-6** Alphanumeric 5 × 7 dot matrix display using two 7489 RAM ICs: (a) circuit; (b) format for constructing desired character.

*Procedure:*

1. Construct the circuit shown in Figure 11-6a.
2. Place $B_{in}$ (pin 1) of the 7493 IC to Sw *I*.
3. Place all data switches *A* through *G* low.
4. Plan the desired character (perhaps the first letter of your first name) on the matrix shown in Figure 11-6b. Each intersection of row lines and column lines indicate a LED. Place solid dots on the LEDs you want to turn on.
5. Activate Sw *I* until the three LED indicators are 0. The memory is now ready to be programmed at location 000.
6. Set data switches *A* through *G* high for all data in column 1 that correspond to the rows. Example: row 1 = *A*, row 2 = *B*, row 3 = *C*, row 4 = *D*, row 5 = *E*, row 6 = *F*, and row 7 = *G*.
7. Activate Sw *H* (WE) low and then high. The column 1 data have been stored.
8. Activate Sw *I* low and then high to set the address inputs to location 001.
9. In a similar manner use steps 6, 7, and 8 to store the data in columns 2 to 5.
10. When all five memory locations have been programmed, remove $B_{in}$ (pin 1) of the 7493 IC from Sw *I* to the Clk.
11. Adjust the clock so that you can see the scanning action of each column as the data on the row inputs turn on the corresponding LEDs.
12. Adjust the clock until your desired character is stationary. If there are any errors in the display, you will have to go back to steps 2 through 10 and reprogram the memory.

*Fill-In Questions:*

1. The columns are scanned with a positive pulse coming from the _____

   _____ IC and the _____ IC.

2. Only five of sixteen memory locations are used, but two memory ICs are

   required because the display has _____ row inputs.

3. The data present at the row inputs to the display that turns on the LEDs

   are in the form of _____.

4. The memory IC 7489-2 uses only three address inputs, three _____

   _____ inputs, and three _____ outputs.

5. The output of pin 12 of the 7404 IC is fed back to the 7493 binary up-

   counter IC as a _____ pulse.

## SECTION 11-5
## INSTANT REVIEW

- A memory consists of memory cells in an array used to store binary data in the form of 1s and 0s. There are address inputs and an address decoder for selecting the various memory locations. It may have data inputs for writing in data and data outputs or sense outputs for reading out data. It will have one or more control inputs to write in and read out data.

- The semiconductor random access memory (RAM) can be written into and read out of any memory location at random. It is a volatile memory.

- The read only memory (ROM) is programmed at the factory and is used only for reading out data. It is a nonvolatile memory.

- A programmable ROM (PROM) is programmed by the user by "blowing" fuse links on the chip. It is considered a nonvolatile memory.

- The erasable PROM (EPROM) has floating gates that are charged during the programming operation and determines the data that are stored. It is a nonvolatile memory that can be erased with ultraviolet light and then reprogrammed.

- The electrical EPROM (EEPROM), also called electrical alterable PROM (EAPROM), can be programmed and erased electrically by the user.

- Memories can be static or dynamic. A static memory has fixed or stationary data such as flip-flops. A dynamic memory refers to a memory that must constantly be "refreshed" to maintain its data, such as a MOS memory using charged capacitances and charge-coupled devices.

- Magnetic memory types are the magnetic core and magnetic bubble. These are nonvolatile memories.

- Auxiliary memories are in the form of magnetic drums, reel-to-reel magnetic tapes, cassettes, normal disks, and floppy disks.

## SECTION 11-6
## EXERCISES II

Perform all the exercises in this section before beginning the next section.

1. Indicate after each type of memory whether it is volatile (V) or nonvolatile (NV).

    a. Semiconductor RAM _____   b. ROM _____

    c. PROM _____   d. EPROM _____

    e. EEPROM _____   f. MOS dynamic RAM _____

    g. Magnetic core RAM _____   h. Magnetic bubble _____

2. Using $256 \times 1$-bit RAM ICs, indicate how many ICs are needed to produce the following memories.

*Type of Memory*                                      *Number of ICs Needed*

**a.** 256 × 8-bit                                    _____

**b.** 256 × 16-bit                                   _____

**c.** 1024 × 8-bit                                   _____

**d.** 4096 × 16-bit                                  _____

**e.** 16,384 × 8-bit                                 _____

**f.** 32,768 × 16-bit                                _____

**g.** 65,536 × 8-bit                                 _____

3. Match the types of memory devices in column A with their proper description in column B.

   *Column A*                                         *Column B*

   **a.** Semiconductor RAM                _____ 1. Has fuse links

   **b.** Semiconductor ROM                _____ 2. A volatile memory

   **c.** PROM                             _____ 3. Programmed at the factory

   **d.** EPROM                            _____ 4. A nonvolatile RAM

   **e.** Magnetic core RAM                _____ 5. Can be erased with ultraviolet light

4. Write RAM or ROM after each statement that properly applies to the corresponding memory device.

   **a.** A nonvolatile memory _____

   **b.** Has both read and write functions _____

   **c.** Considered a permanent memory _____

   **d.** A volatile memory _____

   **e.** Has data inputs _____

   **f.** Has only read input control _____

5. Using Section 11-1e, select the best memory devices that will require the least amount of ICs to construct the following memories.

| | Memory | Number of ICs | Organization Type |
|---|---|---|---|
| Example: | 518 × 8-bit | 4 | 256 × 4-bit |
| a. | 256 × 8-bit | _____ | _____ |
| b. | 1024 × 8-bit | _____ | _____ |
| c. | 4096 × 8-bit | _____ | _____ |
| d. | 16,384 × 16-bit | _____ | _____ |
| e. | 32,768 × 8-bit | _____ | _____ |
| f. | 65,536 × 8-bit | _____ | _____ |

## SECTION 11-7
## TROUBLESHOOTING APPLICATION: TESTING A RAM

A basic functional test for a memory device is that it is capable of storing 1s and 0s in all its memory locations. Figure 11-7 shows how to test the 7489 RAM IC. Input ME is connected to ground so that the LEDs will show the contents of the selected memory location at all times except during the write-in operation (remember that the data will be inverted). The test procedure can be used as follows:

**Figure 11-7**  Testing a 16 × 4-bit word RAM.

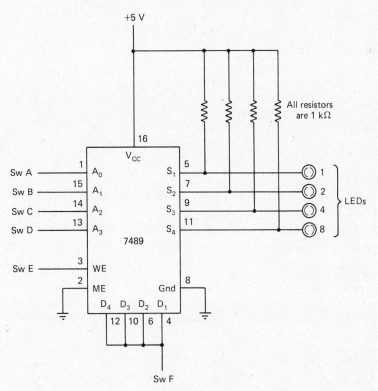

1. Make sure that Sw $E$ is high.
2. Using switches Sw $A$ through Sw $D$, select the desired address.
3. Place Sw $F$ high (all 1s at the data inputs).
4. Move Sw $E$ down and then up. The memory location is loaded with 1s and all LEDs should turn off.
5. Move Sw $F$ low (all 0s at data inputs).
6. Move Sw $E$ down and then up. The memory location is loaded with all 0s, and all LEDs should be on.
7. Repeat steps 2 through 6 for all the other memory locations.

## SECTION 11-8
## SELF-CHECKING QUIZZES

### 11-8a  MEMORY DEVICES: TRUE-FALSE QUIZ

Place a T for true or an F for false to the left of each statement.

_____ 1. A RAM can only be read from.

_____ 2. A PROM is programmed by the user.

_____ 3. The magnetic core memory is volatile.

_____ 4. A semiconductor ROM is a nondestructive readout memory.

_____ 5. An EPROM can be erased with ultraviolet light.

_____ 6. A 256 × 8-bit memory has 256 memory locations (words), each with a word length of 8 bits.

_____ 7. To program a memory means to enter a group of instructions or data in numerical code form into a memory.

_____ 8. "Write in" is the same as to store data into a memory.

_____ 9. An EEPROM can be programmed and erased electrically.

_____ 10. Flip-flops are an example of dynamic memory.

### 11-8b  MEMORY DEVICES: MULTIPLE-CHOICE QUIZ

Circle the correct answer for each question.

1. A memory that can be written into and read out very fast is the:

   a.  RAM          b.  ROM

   c.  PROM         d.  EEPROM

2. A monitor program for a computer would be stored in a:

   a.  RAM          b.  ROM

   c.  magnetic core     d.  magnetic
       RAM                  bubble RAM

3. A specific memory location is accessed by the data present at the:

    a. memory array

    b. data inputs

    c. address inputs

    d. none of the above

4. A $256 \times 4$ memory would have to have the following number of address inputs and data outputs, respectively:

    a. 256 and 4

    b. 64 and 4

    c. 8 and 4

    d. 4 and 8

5. The best type of memory to use for an engineering project would be a:

    a. RAM

    b. ROM

    c. PROM

    d. EPROM

6. Using two IC memories, the word length of a memory can be increased by connecting the:

    a. data lines together (in parallel) and the address inputs separately.

    b. address inputs together (in parallel) and the data lines separately.

    c. address inputs and data lines separately.

    d. none of the above.

7. When two IC memories are connected to increase the number of words in a memory, the total number of:

    a. address inputs remain the same and the data lines increase.

    b. address inputs and data lines increase.

    c. address inputs increase and the data lines remain the same.

    d. address inputs and data lines remain the same.

8. The type of memory that uses a floating gate is the:

    a. RAM

    b. ROM

    c. PROM

    d. EPROM

9. The type of memory that has fusible links is the:

    a. RAM

    b. ROM

    c. PROM

    d. EEPROM

10. Using $256 \times 1$ RAM ICs, the number of ICs required to construct a $256 \times 8$ RAM would be:

    a. 8

    b. 16

    c. 32

    d. 64

**ANSWERS TO EXPERIMENTS AND QUIZZES FOR UNIT 11**

*Experiment 1.*

    (1) 1, 1, 1, 0, 1, 1, 1, A    (2) WE, low    (3) open collector
    (4) limit current    (5) address

*Experiment 2.*

    (1) 7442, 7404    (2) seven    (3) 0s    (4) data, sense    (5) reset

*True-False:*

(1) F   (2) T   (3) F   (4) T   (5) T   (6) T   (7) T   (8) T   (9) T
(10) F

*Multiple-Choice:*

(1) a   (2) b   (3) c   (4) c   (5) d   (6) b   (7) c   (8) d   (9) c
(10) a

# Unit 12

## Digital IC Operation and Specifications

### 12-1a  INTRODUCTION

Most troubleshooting techniques on digital circuits use the logic approach, by looking for 1s and 0s. However, a well-trained technician knows something about the electronic construction and operation of digital ICs. This knowledge is essential for troubleshooting more difficult problems and understanding the use of other components encountered in digital equipment. With the advancement in IC design more circuits or logic elements were fabricated into the same space or area on an IC chip. This technique is called *packing density*. The higher the density, the more circuits there are within a specific area. The various ICs are grouped under a classification, as follows:

- SSI: Small-scale integration, where a single IC package contains less than 12 logic gates.
- MSI: Medium-scale integration, where a single IC package contains more than 12, but fewer than 100 logic gates.
- LSI: Large-scale integration, where a single IC package contains more than 100 logic gates and can go up to thousands of gates.
- VLSI: Very-large-scale integration begins with 1000 logic gates and goes higher. These ICs may contain entire digital systems, such as microprocessors and large memories.
- ULSI: Ultra-large-scale integration, and SLSI, super-large-scale integration, refer to an enormous amount of logic gates, but no specific standard has been set.

307

Of the various groups, SSI and MSI ICs are constructed of TTL and CMOS logic families, whereas LSI and the larger packing density types are generally made of CMOS or MOS technology ICs.

Logic families refer to different function gates produced with the same type of semiconductor construction, such as TTL or CMOS. The main consideration is that circuits belonging to the same logic family can be connected (output to input) directly without any other interface circuitry. Circuits of different families require interface circuits for purposes of buffering or voltage-level changes when connected together.

## 12-1b  TTL FAMILY ICs

### 12-1b.1  Standard TTL IC

*Transistor-transistor logic* (TTL or T²L) refers to the construction of a particular logic family where bipolar transistors are used as switches and are coupled together directly. The standard TTL is the 7400 series. Figure 12-1a shows the NAND gate symbol and its TTL schematic (electronic version) diagram in Figure 12-1b for a single gate of a 7400 quad two-input NAND gate DIP IC.

Transistor $Q_1$ has multiple-emitter inputs with protection diodes that short-out high voltages that could destroy the circuit. The condition of the collector of $Q_1$ controls the state of $Q_2$, which in turn controls the states of $Q_3$ and $Q_4$. The series arrangement of $R_3, Q_3, D_1,$ and $Q_4$ form what is called a *totem-pole* output circuit, similar to the circuit arrangement used in the 555 timer IC introduced in Section 4-1e. When $Q_3$ is conducting, $Q_4$ is off, and vice versa. Diode $D_1$ ensures that only one transistor is conducting at a time.

### 12-1b.1a  TTL NAND Gate Operation When Both Inputs Are High

Figure 12-1c shows the current (conventional) paths for a TTL NAND gate when both inputs are high or a logic 1. The protection diodes are omitted since they have no function in the basic operation. With both input emitters high, $Q_1$ is off and current from $+V_{CC}$ flows through $R_1$ and into the base of $Q_2$, which turns on. Current can now flow from $+V_{CC}$ down through $R_2$, $Q_2$, and $R_4$ to ground. This causes the $V_C$ of $Q_2$ to decrease, which cuts off $Q_3$ (no current flows to the base of $Q_3$). The voltage drop across $R_4$ turns $Q_4$ on, allowing some current to flow into its base. This causes the output to be pulled down toward ground potential or logic 0. If another circuit or gate is connected to the output, it has a path to ground, enabling it to function; this is referred to as *sinking current.*

### 12-1b.1b  TTL NAND Gate Operation When Any One or Both Inputs Are Low

Figure 12-1d shows the different current paths for a TTL NAND gate when any one or both inputs are low or a logic 0. The current flow from $+V_{CC}$ through $R_1$ can now reach ground by one or both of the emitters, which turns $Q_1$ on. The $V_C$ of $Q_1$ decreases, which cuts off $Q_2$ (no current flows to the base of $Q_3$). Since $Q_2$ is off, current flowing from $+V_{CC}$ through $R_2$ flows into the base of $Q_3$ and turns it on. Also, there is no current flow through $R_4$ because $Q_2$ is off, the voltage at the base of $Q_4$ is zero, and $Q_4$ is off. With $Q_3$ being on, current can flow from $+V_{CC}$ through $R_3$, $Q_3$, and

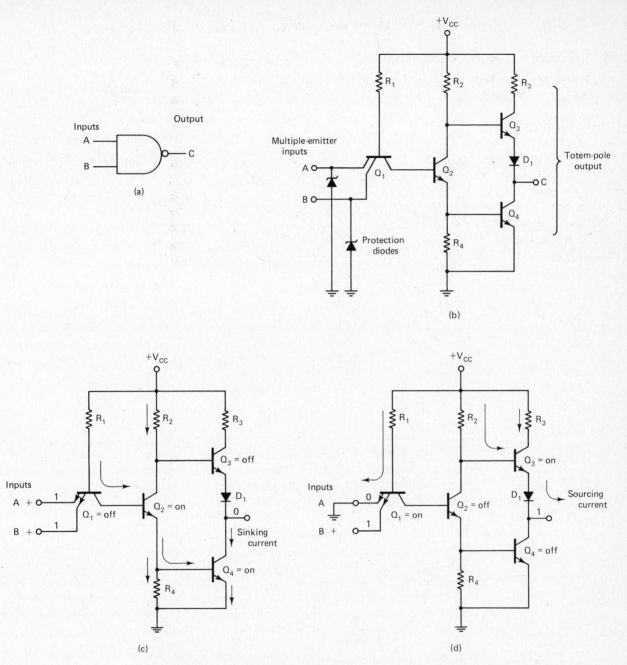

**Figure 12-1** NAND gate TTL circuit: (a) logic symbol; (b) schematic diagram; (c) with all inputs high—gate off; (d) with any one or all inputs low—gate on.

$D_1$ to the output. This causes the output to be pulled up toward $+V_{CC}$ or a logic 1, but because of the voltage drops of $D_1$, $Q_3$, and $R_3$, the output voltage is approximately +3.6 V. With another circuit or gate connected to the output, it has a path to $+V_{CC}$, enabling it to function; this is referred to as *sourcing current*.

Briefly reviewing the operation of a TTL NAND gate, when all emitter inputs are high, the output is near ground potential, and when any one or both emitter inputs are low, the output is high or about +3.6 V. All TTL logic gates function in a similar manner except that they have more or fewer transistors depending on their logic function.

### 12-1b.2 Open-Collector TTL IC

An open-collector TTL NAND gate has components $R_3$, $Q_3$, and $D_1$ omitted; $Q_4$ is the only active device connected to the output. These circuits can be used where the normal TTL totem-pole output configuration is not desired. Such an application is the wired-OR connection that was used with the 7489 RAM IC in Section 11-1b. An output with the open collector requires a pull-up resistor, but can be connected to the outputs of several other circuits. When the output of one open-collector circuit is supposed to be a logic 0, $Q_4$ conducts and the bottom of the pull-up resistor is pulled toward ground, but when its output is supposed to be a logic 1, it does not connect to $+V_{CC}$ and $Q_4$ acts like an open circuit, thereby not interfering with the other outputs connected to it.

These types of ICs are identical pin for pin with their totem-pole output IC counterparts, but have a different designation number and may be listed as O.C. (open collector).

### 12-1b.3 High-Speed TTL IC

The high-speed TTL IC has lower resistor values and a Darlington driver to operate the totem-pole output. Although its switching speed is faster than standard TTL, it also consumes more power. The 74H00 IC is the number designation for the high-speed version of the standard TTL 7400 IC.

### 12-1b.4 Low-Power TTL IC

The low-power TTL IC is basically the same as the standard TTL IC except that the resistor values have been increased, which reduces power consumption. A disadvantage of the low-power TTL IC is that its switching speed is nearly three times slower than a standard TTL IC. The 74L00 IC is the number designation for the low-power version of the standard TTL 7400 IC.

### 12-1b.5 Schottky-Diode Clamped TTL IC

The Schottky-diode clamped TTL IC is the fastest switching circuit of the TTL family. This diode is used to clamp the collector of each transistor to its base, as shown in Figure 12-2a. A normal *pn* junction has a certain storage time from saturation to cutoff because the depletion region must be cleared of carriers (electrons and holes). A Schottky diode has a metal anode and an *n*-type semiconductor material for the cathode, hence no storage time problem, because the electrons flow directly to metal. The Schottky diode reduces the turn on/off time of the transistor and is often indicated as shown in Figure 12-1b. The 74S00 IC is the number designation for the Schottky version of the standard TTL 7400 IC.

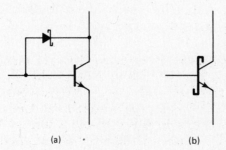

(a)          (b)

**Figure 12-2** Schottky-clamped TTL: (a) Schottky diode between collector and base of transistor; (b) schematic symbol of Schottky-clamped transistor.

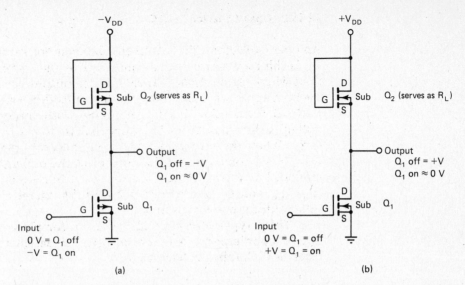

**Figure 12-3**   MOS technology: (a) *p*-channel MOSFET inverter; (b) *n*-channel MOSFET inverter.

### 12-1b.6  Low-Power Schottky TTL IC

The low-power Schottky TTL IC uses Schottky diodes and lower-value resistors to increase the switching speed of a standard TTL IC and have a power consumption of about 2 MW per logic gate. The 74LS00 is the number designation for the low-power Schottky version of the standard TTL 7400 IC.

## 12-1c  MOS TECHNOLOGY AND CMOS LOGIC CIRCUITS

### 12-1c.1  *P*- and *N*- Channel MOS Technology

MOS technology uses enhancement-type MOSFETs in many circuit arrangements. These devices are normally off, which means that when their gate-to-source voltage is 0 V, they are cut off and no current flows through them. This is an ideal situation for digital switching circuits, and they have a high packing density with fewer manufacturing processes than TTL circuits.

   A *p*-channel MOSFET inverter circuit is shown in Figure 12-3a. Transistor $Q_1$ acts as the switch and $Q_2$ is fabricated together with $Q_1$ to serve as the load resistor. When the input is at ground or 0 V, $Q_1$ is off and its drain, or the output, is at about $-V_{DD}$. When the input is at $-V$, $Q_1$ is on and the output is pulled up toward ground or 0 V.

   The *n*-channel MOSFET inverter circuit shown in Figure 12-3b operates in a similar manner. When the input is at 0 V, $Q_1$ is off and the output is at about $+V_{DD}$. When the input is at $+V$, $Q_1$ conducts and the output is at about 0 V.

### 12-1c.2  CMOS Inverter Circuits

A complementary metal-oxide semiconductor (CMOS) circuit is shown in Figure 12-4a. It consists of an *n*-channel and a *p*-channel MOSFET fabricated together with their gates connected together. When the drain of each MOSFET is connected together, it forms an inverter circuit. If the input voltage to the gates is 0 V, the *n*-channel MOSFET is off and the *p*-channel MOSFET is on. This allows the output to be pulled up toward $+V_{DD}$, as

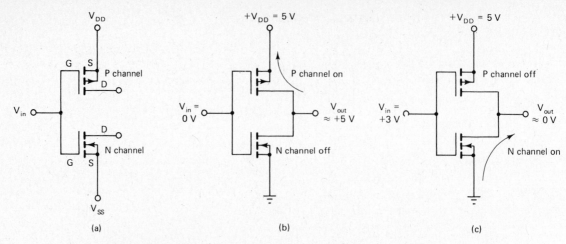

**Figure 12-4** Complementary metal-oxide semiconductor (CMOS): (a) schematic diagram; (b) *p* channel on; (c) *n* channel on. (From F. Hughes, *Basic Electronics: Theory and Experimentation,* Prentice-Hall, Englewood Cliffs, N.J., © 1984, Fig. 5-10, p. 123. Reprinted with permission.)

shown in Figure 12-4b. When the input voltage is +3 V, the *p*-channel MOSFET is off and the *n*-channel MOSFET is on, which pulls the output toward ground level, as shown in Figure 12-4c.

### 12-1c.3  CMOS NAND Gate Circuit

Using the same idea as the CMOS inverter circuit, a CMOS NAND gate is constructed as shown in Figure 12-5. When both inputs are at +V, $Q_1$ and $Q_2$ are cut off, while $Q_3$ and $Q_4$ are turned on and pull the output toward ground level. If input A goes to 0 V, $Q_3$ turns off, opening the output from ground, but $Q_1$ turns on and pulls the output up toward $+V_{DD}$. If input B goes to 0 V, $Q_4$ turns off, opening the output from ground while $Q_2$ turns on and pulls the output up toward $+V_{DD}$. When both inputs are 0 V, both $Q_3$ and $Q_4$ are cut off and both $Q_1$ and $Q_2$ are on, with the output pulled up to approximately the $+V_{DD}$ level.

**Figure 12-5**  CMOS NAND gate: (a) logic symbol; (b) schematic diagram.

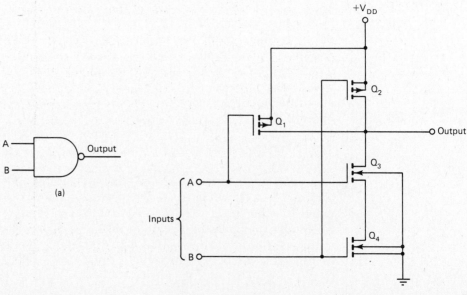

### 12-1c.4  CMOS Logic Voltage Levels

CMOS ICs can operate with a $+V_{DD}$ (or $+V_{CC}$) voltage of from $+3$ to $+15$ V. However, the logic voltage levels required for CMOS are higher than for TTL ICs. The voltage percentages for CMOS are:

Logic 1 = a minimum of 75% of $+V_{CC}$

Logic 0 = a maximum of 25% of $+V_{CC}$

Undefined region = voltage levels between these limits

For example, with a $+V_{CC}$ of $+5$ V, the TTL requires $+2$ V to represent a logic 1, whereas CMOS would require $+3.75$ V ($V_{CC} \times 0.75 = 3.75$). Similarly, the maximum voltage level for a logic 0 for TTL is $+0.8$ V, and for CMOS it would be $+1.25$ V ($V_{CC} \times 0.25 = 1.25$). If CMOS ICs are used with a $+10$-V power supply, the logic 1 would have to be at least $+7.5$ V ($V_{CC} \times 0.75 = 7.5$), and a logic 0 would have to be a maximum of $+2.5$ V ($V_{CC} \times 0.25 = 2.5$).

### 12-1c.5  CMOS IC Features

There are three types of CMOS ICs. The older A-type is unbuffered and resembles those shown in Figures 12-4 and 12-5. The buffered B-type consists of a basic A-type gate structure followed by two CMOS inverter stages. The B-type provides an equal drive output for high and low conditions and as a result of improved fabrication techniques is slightly faster than the A-type. The third type, the silicon-on-sapphire (SOS), is about 10 times faster than the B-type with less power consumption, but is more expensive. Compared to TTL ICs, CMOS ICs have:

1. Slower switching speeds
2. Lower power consumption
3. Lower input power requirements (no current flow)
4. Better noise immunity
5. No switching voltage transients
6. No current spikes
7. A large fan-out capability of about 50 when used with other CMOS ICs

There are two widely used CMOS IC series. One is the 74C00 series, which has many ICs that are pin-for-pin compatible with the TTL 7400 series. The other type is the popular 4000 CMOS series, some of which are compatible with TTL, but its number identification system is not the same as the 74C00 series. Table 12-1 shows this comparison statistically.

Precautionary measures should be used before handling all MOS technology devices, even if they have built-in protection diodes. Static voltage charges built up on your body, and/or materials such as plastic, could rupture the thin oxide layer between the gates and channels of the MOSFETs, thus destroying the device. The following procedures are recommended when working with MOS devices:

1. Ground yourself to the circuit being worked on by using special grounding straps or a clip lead attached to your wristwatch band and ground.

**TABLE 12-1**

General Comparison TTL and CMOS IC Family Types, Based on the NAND Gate, with a $V_{CC}$ of +5 V

| Logic Family Type | Average Power Dissipation per Gate (MW) | Average Propagation Delay per Gate (ns) | Clock Rate or Frequency Input (MHz) |
|---|---|---|---|
| TTL standard | 10 | 11 | 35 |
| TTL high-speed | 22 | 6 | 50 |
| TTL low-power | 1 | 45 | 3 |
| TTL Schottky | 20 | 5 | 125 |
| TTL low-power, Schottky | 2 | 8 | 45 |
| CMOS 74C00 | Depends on frequency, | 90 | 2 |
| CMOS A-type | but much lower | 150 | 1 |
| CMOS B-type | than TTL | 180 | 2.5 |

2. Ground the tips of the soldering iron or any heat device used.

3. Do not use a soldering gun because the ac voltage radiation may damage the MOS devices.

4. Carefully remove the MOS devices from the nonstatic foam or plastic package that they come in and quickly place them in the circuit.

5. Use the least amount of time and heat at each solder connection.

### 12-1d  THREE-STATE LOGIC CIRCUITS

With complex digital circuits, the outputs of gates and flip-flops are connected to common lines called *data transfer buses*. If the output of a particular gate is high while the other gates are low, a short-circuit effect results, which impairs circuit operation and may cause excessive current flow, possibly damaging some of the gates. To overcome this problem with common busing lines, *three-state logic* (TSL), also called *three-state TTL* or *tri-state logic*, was developed.

A control input is added to the logic device as shown in Figure 12-6a. When this control input is a logic 0, the device is enabled and performs normally, but when the control input is 1, the device appears as a high resistance output or open circuit and essentially is disconnected from the bus line.

A similar arrangement is used for CMOS ICs, as shown in Figure 12-6b. In this case, a CMOS *transmission gate* is connected to the output of a CMOS logic device. The transmission gate will allow data to pass in either direction and is also referred to as a *bilateral switch*. When the control input is a logic 0, the switch is off and presents a high resistance or open circuit to the output of the logic device. When the control input is a logic 1, the switch is on and allows data to pass to the output.

Figure 12-6c shows an example of how three-state logic devices may be used with a data bus line. If gate *A* is supposed to send data to gate *C*, the select input (control inputs) to these gates will be logic 0 and the select inputs of the other gates will be logic 1. At another time the reverse may be true, and gate *B* could send data to gate *D*. In any case, only one sending and one receiving gate is on at a time and there is no interference from the other gates.

### 12-1e  OTHER TYPES OF LOGIC FAMILIES

*Emitter-coupled logic* (ECL) is a bipolar transistor circuit based on a nonsaturating current-switching circuit similar to the analog differential am-

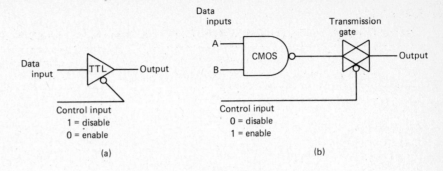

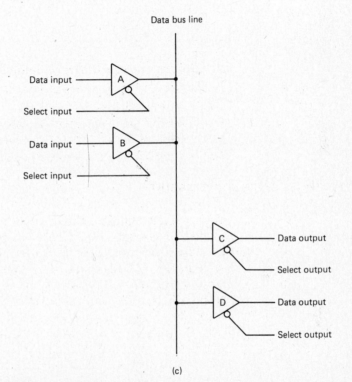

**Figure 12-6** Three-state gates:
(a) TTL three-state driver;
(b) CMOS NAND gate with transmission gate; (c) send/receive three-state control data bus line.

plifier. The speed of ECL-type circuits is 1-ns propagation with a power consumption of 60 MW per gate and a 2-ns gate with a power loss of 25 MW.

The ECL logic family is also different from the other two types of logic families in that it uses a −5-V power supply with the maximum limit for a logic 1 at −0.8 V and the maximum limit for a logic 0 at −1.6 V. The main advantage of ECL is its high operating speed.

*Integrated injection logic* ($I^2L$), also called *merged transistor logic* (MTL), is a bipolar transistor circuit where the injection current (current flow through the circuit) can be programmed or set with an external resistor which varies the propagation delay and power dissipation of the IC over a very wide range. Also, its fabricating techniques allow greater packing density, which competes with MOS devices, but with the higher TTL switching speeds.

## 12-1f  DIP IC IDENTIFICATION

A large number of digital integrated circuits are mounted in dual-in-line packages (DIP) as shown in Figure 12-7. There is some standardization for identifying the contents of these packages. The first two letters indicate the

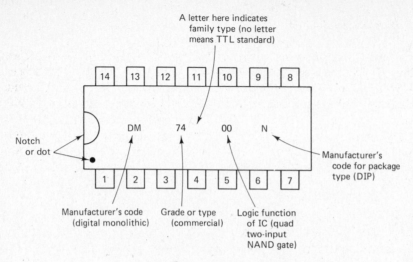

Figure 12-7   IC identification.

manufacturer's code (DM stands for digital monolithic). The next two numbers (74), indicate that it belongs to the 7400 series ICs, which are suitable for industrial-grade applications or those that require a temperature operating range of 0 to +70°C. If these numbers are 54, the IC belongs to the military grade, with a temperature operating range from −55 to +125°C. A letter following these numbers indicates which specific logic family the IC belongs to (i.e., H = high-speed TTL, L = low-power TTL, S = Schottky TTL, LS = low-power Schottky TTL, and C for CMOS). No letter indicates that it is standard TTL. The next two numbers show which logic function the package contains (i.e., AND, OR, NAND, NOR, etc.). The last letter is the manufacturer's code for the type of package (i.e., N = dual-in-line).

To identify the pins, you have to place the notch or dot on one end of the package to the left. The pins are numbered counterclockwise beginning with the lower-most-left pin as number 1.

## 12-1g   GENERAL DIGITAL IC SPECIFICATIONS

### 12-1g.1   Fan-Out

The *fan-out* of a logic gate or circuit is the number of other gates that can be connected to it without causing such a loading problem that circuit operation becomes faulty or impaired. As any logic gate is loaded down it becomes more susceptible to noise and its 0 and 1 logic voltage levels can become affected. The standard fan-out of a typical TTL logic device is 10, which means that the manufacturer guarantees that it can drive 10 other logic devices satisfactorily. This factor varies with the TTL subfamilies and when they are interconnected. The standard fan-out for CMOS is 50.

### 12-1g.2   Propagation Delay

*Propagation delay* is the speed at which a logic circuit switches its output from low to high, or vice versa, in response to an input signal. This speed is the time it takes a gate, after a fast change at the input, to make a resulting change at the output and is usually expressed in microseconds (millionths of a second) and in nanoseconds (billionths of a second). It is ideal for a logic circuit to have the fastest possible propagation delay, so as to be able to process a large amount of information in a short time. (Refer to Figure 4-5a and b to get a visual idea of propagation delay.)

### 12-1g.3  Power Dissipation

The *power dissipation* of any electrical device consists of the waste and scattering of electrical energy in the form of heat generated by both the currents and voltages involved. The power dissipation of digital ICs is usually expressed in milliwatts (thousandths of a watt), and an ideal circuit would have a very low rating. Because heat is a by-product of wasted power, it must quickly be removed from the IC to the surrounding air. When circuits are crowded together in an IC, it is more difficult for the heat to get out of the silicon into the air; therefore, the circuits may become too hot to operate properly.

When the propagation delay of an IC is short, the IC works faster (and more), thereby generating more heat, and its power dissipation is usually higher. This is a general trade-off factor when digital circuits are designed (i.e., faster-operating circuits consume more power and slower-operating circuits consume less power).

### 12-1g.4  Noise Margin

The *noise margin* or *noise immunity* of a digital circuit is its ability to ignore extraneous voltage pulses (noise) that occur between the threshold voltages set for a logic 0 and 1. The noise margin for TTL ICs is set at approximately 1 V, which means that a noise pulse greater than this would be in the forbidden region and may unwantingly operate a logic circuit. (Refer to Figure 4-5d to get a visual ideal of noise margin.) The noise margin for CMOS ICs is much higher, at about 45% of $V_{CC}$. Assuming that $V_{CC}$ is +5 V, TTL has a noise margin of 20% or 1 V and CMOS at 45% or 2.25 V.

## 12-1h  INTERFACING LOGIC FAMILIES

Interfacing in electronics means a point or boundary at which a transition is performed to make different circuits or systems compatible (operate together) with one another in terms of speed, modes of operation, power levels (voltage and current), and so on. The interfacing of digital circuits such as TTL and CMOS usually requires other electronic components. There are many possible ways to interface circuits, and a few will be shown to give an idea of how this is accomplished.

### 12-1h.1  Interfacing TTL to CMOS

Interfacing a TTL IC to a CMOS IC using the same +5-V power supply may only require a pull-up resistor, as shown in Figure 12-8a. The pull-up resistor is used to ensure that a logic 1 will be of sufficient voltage amplitude to operate the CMOS circuit properly. Remember, this logic level is greater for CMOS than for a TTL logic 1.

**Figure 12-8**  Interfacing TTL to CMOS: (a) with the same +5-V power supply; (b) with two different power supplies.

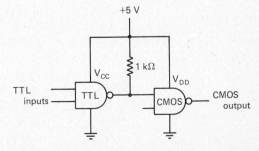

*(continued)*

**Figure 12-8 (cont.)**

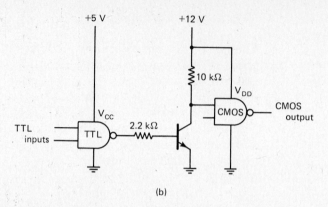

(b)

When two different power supplies are used, a transistor circuit may be used to isolate or buffer the TTL IC from the CMOS IC and also be able to drive the CMOS IC input, as shown in Figure 12-8b.

### 12-1h.2  Interfacing CMOS to TTL

To interface CMOS ICs to TTL ICs it is probably easier to use a special buffer/converter IC such as the CD4050. With the same +5-V power supply, all three ICs are connected to the same voltage, as shown in Figure 12-9a. With different power supply voltages, the buffer/converter IC is connected to the voltage used with the TTL IC, as shown in Figure 12-9b.

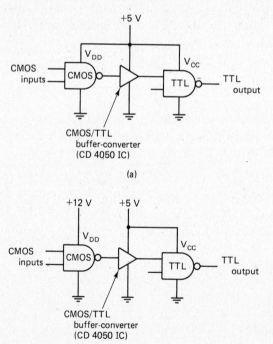

**Figure 12-9**  Interfacing CMOS to TTL: (a) with the same +5-V power supply; (b) with two different power supplies.

## 12-1i  POWER SUPPLY DECOUPLING

Current spikes are generated by the on and off switching operations of TTL ICs. These spikes can affect the power supply of a digital system and actually appear as voltage spikes on the $+V_{CC}$ line throughout the system. Needless to say, this situation can affect the operation of other digital circuits. The

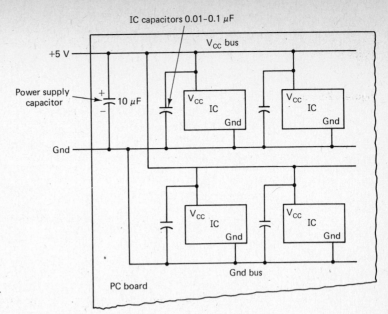

**Figure 12-10** Decoupling capacitors on PC board.

unwanted voltage spikes are reduced or eliminated by the use of capacitors across the $+V_{CC}$ and ground lines. By this method the spikes are filtered out or decoupled from the power supply.

Usually, a 10-$\mu$F decoupling capacitor is placed across the power lines where they enter a printed circuit (PC) board, as shown in Figure 12-10. Smaller, 0.01- to 0.1-$\mu$F decoupling capacitors are placed across the power lines where they connect to each IC, also shown in Figure 12-10. At least one decoupling capacitor should be used for every four SSI ICs, every two MSI ICs, and any IC that is over 3 inches from the nearest decoupling capacitor. The capacitor leads should be kept as short as possible to minimize any effects of inductive reactance, which may tend to offset the decoupling capacitor's function.

**SECTION 12-2**
**EXERCISES I**

Perform all the exercises in this section before beginning the next section.

1. Draw the schematic diagram for a TTL NAND gate that would be found in a standard 7400 IC (label all components).

2. Referring to Exercise 1, list the transistors that are on and those that are off when both inputs are at a logic 1.

| Transistors | |
|-----|-----|
| ON | OFF |

3. Referring to Exercise 1, list the transistors that are on and those that are off when either or both inputs go to a logic 0.

| Transistors | |
|-------------|-----|
| ON | OFF |

4. Draw the schematic diagram for a standard TTL NAND gate with an open collector.

5. Draw the schematic diagram of a CMOS inverter as shown in Figure 12-4 (connect the drains together).

6. Referring to Exercise 5, which MOSFET is on when the output is a logic 1?

_____-channel MOSFET

7. Referring to Exercise 5, which MOSFET is on when the output is a logic 0?

_____-channel MOSFET

8. Draw the logic sysmbol for a three-state TTL NAND gate.

9. Draw a simple method of interfacing TTL to CMOS using the same power supply.

10. Draw a simple method of interfacing CMOS to TTL when the power supplies are different.

11. List the precautions used when working with MOS devices.

   1. _____

   2. _____

   3. _____

   4. _____

   5. _____

12. List the entire identification of the following ICs (refer to Units 2 and 3 for device types).

   | ICs | Identification |
   |-----|----------------|
   | a. DM 74L00 | _____ |
   | b. DM 7402N | _____ |
   | c. DM 74S08J | _____ |
   | d. DM 74H32N | _____ |
   | e. DM 54LS86J | _____ |
   | f. MM 74C04 | _____ |

## SECTION 12-3
## DEFINITION EXERCISES

Give a brief description of each of the following terms.

1. SSI _____

_____

_____

2. MSI _____

_____

_____

3. LSI _____

_____

_____

4. VLSI _____

_____

_____

5. ULSI _____

_____

_____

6. SLSI _____

_____

_____

7. TTL ($T^2L$) _____

_____

_____

8. CMOS _____

_____

_____

9. DIP _____

_____

_____

10. Source _____

_____

_____

11. Sink _____

_____

_____

12. Schottky diode _____

_____

_____

13. ECL _____

_____

_____

14. $I^2L$ _____

_____

_____

15. MTL _____

_____

_____

16. Open collector _____

_____

_____

17. Three-state logic _____

_____

_____

18. Fan-out _____

_____

_____

19. Propagation delay _____

_____

_____

20. Power dissipation _____

_____

_____

21. Noise margin _____

_____

_____

22. Noise immunity _____

_____

_____

23. Interfacing _____

_____

_____

24. Decoupling _____

_____

_____

25. Data bus line _____

_____

_____

26. Packing density _____

_____

_____

27. Totem-pole output _____

_____

_____

28. Multiple-emitter input _____

_____

_____

## SECTION 12-4
## EXPERIMENTS

### EXPERIMENT 1. COMPARISON OF TTL AND CMOS LOGIC 1 VOLTAGE LEVELS

*Objective:*

To demonstrate the difference in voltage levels needed to operate TTL ICs and CMOS ICs.

*Introduction:*

In the first part of this experiment a TTL NAND gate is tested for the logic 1 voltage levels needed to turn it off. The second part of the experiment tests the same conditions for a CMOS NAND gate, but includes varying values of power supply voltage. The values of voltage used are well within the manufacturer's guaranteed minimum values, although the circuit may respond to lesser values.

*Materials Needed:*

1 Universal breadboard

1 Digital or standard voltmeter

1 Variable power supply of 0 to +15 V dc

1 7400 quad two-input NAND gate DIP IC (TTL)

1 CD 4011 quad two-input NAND gate DIP IC (CMOS)

2 10-kΩ potentiometers (linear taper)
Several leads or hookup wires

*Procedure:*

*TTL*

1. Construct the circuit shown in Figure 12-11a. Be careful not to exceed the +5-V power supply level, or the 7400 TTL IC may be destroyed.

2. Using the voltmeter and potentiometers, adjust the input voltages given in the second and third columns of the TTL table shown in Figure 12-11b.

3. Using the voltmeter, measure $V_{out}$ and record its value in the fourth column of the TTL table.

4. Using the formula given, calculate the percentage of voltage needed for a logic 1 with the data in the third and fourth columns. Record this percentage in the fifth column of the TTL table.

5. In the sixth column of the TTL table, indicate if the gate is on or off (on when the output is 1 and off when the output is 0). Perform the same procedures for both lines.

**Figure 12-11** Comparison of voltage levels for TTL and CMOS ICs: (a) TTL IC circuit; (b) TTL table; (c) CMOS IC circuit; (d) CMOS table.

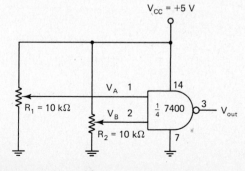

| $V_{CC}$ | $V_{Ain}$ | $V_{Bin}$ | $V_{out}$ | Logic 1 percentage % = $V_{in}/V_{CC}$ | Gate condition (on/off) |
|---|---|---|---|---|---|
| +5 V | +1 V | +1 V | | | |
| +5 V | +2 V | +2 V | | | |

(a)

(b)

*(continued)*

**Figure 12-11 (cont.)**

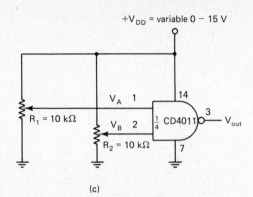

(c)

| $V_{DD}$ | $V_{Ain}$ | $V_{Bin}$ | $V_{out}$ | Logic 1 percentage % = $V_{in}/V_{DD}$ | Gate condition (on/off) |
|---|---|---|---|---|---|
| +5 V | +2 V | +2 V | | | |
| +5 V | +3.5 V | +3.5 V | | | |
| +12 V | +3.5 V | +3.5 V | | | |
| +12 V | +9 V | +9 V | | | |
| +15 V | +8 V | +8 V | | | |
| +15 V | +10.5 V | +10.5 V | | | |

(d)

*CMOS*

6. Construct the circuit shown in Figure 12-11c.
7. Using the CMOS table shown in Figure 12-11d, adjust $V_{DD}$, $V_{Ain}$, and $V_{Bin}$ accordingly.
8. Perform the same procedures for the entire table as was done for the TTL circuit.

*Fill-In Questions:*

1. The guaranteed percentage of $V_{CC}$ needed at each input for a logic 1 to the TTL NAND gate to turn it off is _____%.

2. With a $V_{DD}$ of +5 V and the same voltage level of inputs to the TTL circuit applied to the CMOS circuit, the NAND gate _____ ____ _____ turn off.

3. The guaranteed percentage of $V_{DD}$ for a logic 1 that turned off the CMOS NAND gate was _____%.

4. With a $V_{DD}$ of +12 V, the CMOS NAND gate did not turn off with _____ V applied to its inputs, but did turn off when this voltage was _____ V.

5. The percentage of voltage shown in the CMOS table that turned off the

   CMOS NAND gate with a $V_{DD}$ of +15 V was _____%.

**SECTION 12-5
INSTANT REVIEW**

- TTL digital ICs use bipolar transistor construction. They use a fixed power supply of +5 V and require a +2-V supply for logic 1 operation. TTL ICs have faster switching speeds than CMOS ICs, but consume more power.
- An open-collector TTL IC has only a sinking transistor connected to the output and requires a pull-up resistor or other device for proper operation.
- Other types of TTL ICs are:

  | | |
  |---|---|
  | TTL high speed | 74HXX series |
  | TTL low power | 74LXX series |
  | TTL Schottky | 74SXX series |
  | TTL low-power Schottky | 74LSXX series |

- CMOS digital ICs use MOSFET construction. They operate over a wider range of power supply voltages (+3 to +18 V) and require 75% of $V_{DD}$ for a logic 1 operation. CMOS ICs consume much less power than do TTL ICs, take less space in a package, but have slower switching speeds.
- Types of CMOS ICs are:

  | | |
  |---|---|
  | CMOS pin-for-pin compatible with TTL | |
  | CMOS type-A buffered } | 74CXX series |
  | CMOS type-B buffered } | CD40XX series |

- Three-state logic ICs have a control input that causes the output to operate normally or appear as an open circuit.
- Newer logic families with faster switching speeds and low power consumption are ECL and I²L.
- Fan-out is the number of other gates a single output of a gate can drive safely.
- Propagation delay is the switching time of a digital circuit.
- Power dissipation is the amount of power used by a digital circuit.
- The noise margin of a digital circuit is the guaranteed voltage level for a logic 1 to which the circuit will respond.
- Since the voltage and current requirements are different for TTL and CMOS ICs, a form of interfacing is needed between the two types when they are used together.
- Decoupling digital ICs is a means of using capacitors across the power supply lines to filter out voltage transients (spikes) as a result of the switching operation of the ICs.
- Sourcing current (conventional flow) flows from a larger magnitude voltage level.
- Sinking current (conventional flow) flows to ground.

**SECTION 12-6**
**EXERCISES II**

Perform all the exercises in this section before beginning the next section.

1. Draw the schematic diagram for a CMOS NAND gate.

2. Referring to Exercise 1, indicate which transistors are on and off for the following input conditions:

| Input | | Transistors | | | |
|---|---|---|---|---|---|
| A | B | $Q_1$ | $Q_2$ | $Q_3$ | $Q_4$ |
| 0 | 0 | | | | |
| 0 | 1 | | | | |
| 1 | 0 | | | | |
| 1 | 1 | | | | |

3. Draw the shape of a DIP IC and label the identification printed on it (refer to Figure 12-7).

4. Draw a method of interfacing TTL to CMOS with two different power supplies.

5. Draw a method of interfacing CMOS to TTL using the same power supply.

6. Draw a section of a PC board and indicate the use of capacitors for decoupling digital circuits from the power supply.

7. The following statements provide a comparison of TTL and CMOS ICs. After each statement, indicate if it applies to TTL or CMOS.

   a. Has a faster switching speed. _____

   b. Has lower power consumption. _____

   c. Has a fan-out of 10. _____

   d. Has better noise immunity. _____

   e. Can use a wider range of power supply values. _____

   f. Has a slower switching speed. _____

   g. Has a fan-out of about 50. _____

   h. Generally has a higher power consumption. _____

   i. Has a fixed +5-V power supply. _____

   j. Requires a 40% level of the power supply to operate at a logic 1. _____

   k. Special care in handling must be used. _____

   l. Requires a 70% level of the power supply to operate at a logic 1. _____

## SECTION 12-7
## TROUBLESHOOTING APPLICATION: A BINARY COUNTING SYSTEM

The troubleshooting applications of previous units dealt with the testing of single or discrete digital components. In this case, a binary MOD-10 counter is connected to a BCD-to-seven-segment decoder, which in turn is connected to a seven-segment LED display as shown in Figure 12-12. Initially, the system is constructed and then tested to make sure that digits 0 through 9 will light up on the display. Next, the counter is set to zero and a problem is introduced into the system. Then the LED display is observed to notice the trouble symptom. To begin troubleshooting, the first thing to establish is

**Figure 12-12** Troubleshooting a binary counting system.

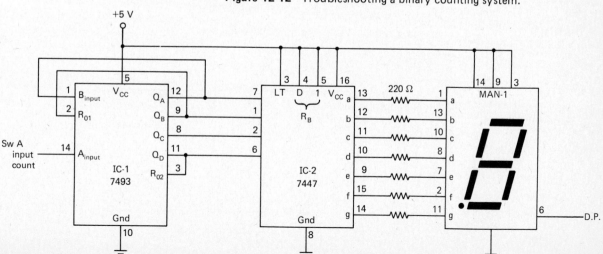

the normal condition for the LED display with the known contents of the counter. Second, the incorrect condition is established. Finally, a troubleshooting path is selected, and then testing can begin with a logic probe. Testing may start at the input to a system, but for the practice of circuit tracing it will begin at the LED display.

The LED display is the common-anode type; therefore, the 220-$\Omega$ series current limiting resistors are connected to the cathodes. The decoder IC-2 sinks the current to light the LEDs. When this condition exists, there is a voltage drop across the resistors and a logic 1 will appear at the display pins. When there is no current flow through the resistors, a logic 1 will appear at the display pins because of the open-circuit condition. Therefore, logic 1s will appear on all the display pins all the time.

Efficient troubleshooting results in constant reference being made from the logic diagram to the actual circuit. The following example will help in understanding how the procedure is performed.

**Example:**

| Trouble Condition | Problem Component | Display Symptom |
|---|---|---|
| Open connection | IC-1, pin 8 | |

Normal condition: If the counter is at 0, segments a, b, c, d, e, and f of the display should be on.
Incorrect condition: Segments a, d, and e are off.
Select troubleshooting path: Segment a.

| Testing Sequence | Test Point | Logic Result | Conclusion | Comment |
|---|---|---|---|---|
| 1 | Display pin 1 | 1 | Normal | Should always be 1 |
| 2 | IC-2 pin 13 | 1 | Normal | For segment a being off |
| 3 | IC-2 pin 7 | 0 | Normal | When counter is 0 |
| 4 | IC-2 pin 1 | 0 | Normal | When counter is 0 |
| 5 | IC-2 pin 2 | 1 | Not normal | For 0, but normal for 4 |
| 6 | IC-2 pin 6 | 0 | Normal | |
| 7 | IC-1 pin 8 | 0 | Normal | When counter is 0 |
| 8 | IC-1 pin 12 | 0 | Normal | When counter is 0 |
| 9 | IC-1 pin 9 | 0 | Normal | When counter is 0 |
| 10 | IC-1 pin 11 | 0 | Normal | When counter is 0 |

Trouble location: Between IC-1, pin 8, and IC-2, pin 2.

Perform the following three troubleshooting problems in a similar manner.

**Troubleshooting Problem 1:**

| Trouble Condition | Problem Component | Display Symptom |
|---|---|---|
| Open connection | IC-1, pin 11 | |

Normal condition:

Incorrect condition:

Select troubleshooting path:

| Testing Sequence | Test Point | Logic Result | Conclusion | Comment |
|---|---|---|---|---|
| | | | | |

## Troubleshooting Problem 2:

| Trouble Condition | Problem Component | Display Symptom |
|---|---|---|
| Open connection | IC-2, pin 15 | |
| Normal condition: | | |
| Incorrect condition: | | |
| Select troubleshooting path: | | |

| Testing Sequence | Test Point | Logic Result | Conclusion | Comment |
|---|---|---|---|---|
| | | | | |

**Troubleshooting Problem 3:**

| Trouble Condition | Problem Component | Display Symptom |
|---|---|---|
| Open connection | Display, pin 10 | ⌐ ⌐ |
| Normal condition: | | |
| Incorrect condition: | | |
| Select troubleshooting path: | | |

| Testing Sequence | Test Point | Logic Result | Conclusion | Comment |
|---|---|---|---|---|

## SECTION 12-8
## SELF-CHECKING QUIZZES

### 12-8a  DIGITAL IC OPERATION: TRUE-FALSE QUIZ

Place a T for true or an F for false to the left of each statement.

_____ 1. CMOS digital ICs have faster switching speeds than those of TTL ICs.

_____ 2. A pull-up resistor is usually required with an open-collector type IC.

_____ 3. Three-state logic ICs allow the outputs of the circuits to operate normally or acts like a short to ground.

_____ 4. TTL digital ICs consume less power than do CMOS ICs.

_____ 5. An IC with the designation 74LS00 means a TTL low-speed quad two-input NAND gate.

_____ 6. The output of a single CMOS gate can be connected to the inputs of 25 other CMOS gates.

_____ 7. A 10-$\mu$F decoupling capacitor is usually used where the power supply lines enter a PC board.

_____ 8. In a CMOS circuit using a $V_{DD}$ of +12 V, a logic 1 would have to be + 6.5 V.

_____ 9. TTL ICs and CMOS ICs can be connected together directly.

_____ 10. The fan-out for a TTL gate is 20.

## 12-8b  DIGITAL IC OPERATION: MULTIPLE-CHOICE QUIZ

Circle the correct answer for each question.

1. When the output of a digital IC is low, the current that flow is referred to as:

   a. minimum       b. maximum

   c. sourcing       d. sinking

2. The designation 74C00 indicates that the device is:

   a. TTL standard    b. TTL low power

   c. CMOS         d. TTL capacitor-coupled

3. A capacitor value that would be used across an IC for decoupling purposes is:

   a. 10 $\mu$F       b. 0.05 $\mu$F

   c. 0.001 $\mu$F      d. 100 $\mu$F

4. The type of logic circuits used to connect to a data bus line would normally be:

   a. three-state logic   b. totem-pole output

   c. TTL          d. CMOS

5. An IC package containing 90 logic gates would be classified as:

   a. SSI         b. MSI

   c. LSI         d. VLSI

6. The logic family ICs that require special precautionary measures when handling are:

   a. TTL         b. CMOS

   c. ECL         d. $I^2L$

7. The time it takes the output of a digital circuit to respond to a change at the inputs is called:

   a. noise margin     b. fan-out

   c. propagation delay   d. power dissipation

8. The method used to connect TTL ICs to CMOS ICs is called:

   a. buffering      b. current sourcing

   c. decoupling      d. interfacing

9. The NAND gate IC to use in a circuit requiring the least power consumption is:

   a. 74S00        b. 74L00

   c. 7400         d. 74C00

10. The inverter IC to use in a circuit requiring the fastest switching speed is:

   a. 74H04        b. 74S04

   c. 7404         d. 74C04

**ANSWERS TO EXPERIMENTS AND QUIZZES FOR UNIT 12**

*Experiment 1.*

       **(1)** 40    **(2)** will not    **(3)** 70    **(4)** 3.5, 9    **(5)** 70

*True-False:*

       **(1)** F    **(2)** T    **(3)** F    **(4)** F    **(5)** F    **(6)** T    **(7)** T    **(8)** F    **(9)** F
       **(10)** F

*Multiple-Choice:*

       **(1)** d    **(2)** c    **(3)** b    **(4)** a    **(5)** b    **(6)** b    **(7)** c    **(8)** d    **(9)** d
       **(10)** b

# Appendix

# Basic Digital Test Equipment

This appendix provides information for constructing a basic logic probe and a logic pulser. Your instructor may give extra credit points for completion of these projects.

## GENERAL CONSTRUCTION SUGGESTIONS

These two basic projects are constructed on perforated board (perfboard) with holes 0.001 inch on center. The component leads are pushed through the holes in the perfboard and then bent over to hold the component in place. Most of the component leads are long enough to make direct soldering connections. It is recommended that an IC socket be used to assemble the projects and then carefully insert the IC for operation, although the IC leads can be soldered directly if care is used.

The following list of tools and items are used to assemble the projects:

1 Hacksaw or electric jig saw (for cutting perfboard)
1 Electric hand drill and assorted drill bits
1 Small file
1 Small screwdriver
1 Long-nose pliers
1 Diagonal cutters
1 Wire stripper
1 37½- to 50-W soldering iron
1 Soldering holder and sponge
1 Thin resin-core (60/40 tin-nickel alloy) solder, about 3 feet long
  Assorted hardware: screws, nuts, bolts, washers, No. 30 gauge stranded-insulated wire, etc.

It is recommended that you gather all the parts before beginning actual construction on the projects.

## TTL/CMOS LOGIC PROBE

### LOGIC PROBE PARTS LIST

1   555 timer mini-DIP IC
1   8-pin mini-DIP IC socket
1   1N4001 diode or equivalent ($D_1$)
1   1N5231 zener diode or equivalent with $V_z = 5.1$ V ($Z_1$)
1   Red LED with $V_F \approx 2$ V (LED$_1$)
1   Green LED with $V_F \approx 2$ V (LED$_2$)
1   100-$\Omega$ resistor at 0.5 W ($R_1$)
1   220-$\Omega$ resistor at 0.25 W ($R_2$)
2   330-$\Omega$ resistors at 0.25 W ($R_4$ and $R_5$)
1   1-k$\Omega$ resistor at 0.25 W ($R_3$)
1   0.01-$\mu$F capacitor at 25 WV dc ($C_1$)
1   Perfboard (approximately $1 \times 3$ inches)
1   Clear plastic medicine bottle that will accommodate the perfboard
1   Aluminum probe or finishing nail
1   No. 22 gauge stranded-insulated wire (leads)
1   Red mini-alligator clip
1   Black mini-alligator clip

### LOGIC PROBE THEORY OF OPERATION

The TTL/CMOS logic probe schematic diagram is shown in Figure A-1a and a parts layout is shown in Figure A-1b. Diode $D_1$ is used as a protection device in case the power supply leads are incorrectly attached to the circuit being tested. When the logic probe is used on TTL circuits with a +5-V power supply, the 555 timer operates at a little less voltage and the LED indicators will respond appropriately to the logic condition at the probe input. When the logic probe is used on CMOS circuits with a higher voltage power supply (up to about 15 V), the zener diode ($Z_1$) operates and supplies about +5 V to the 555 timer IC and LED indicators. The proper LED indicator will also light, depending on the logic condition at the probe input.

### LOGIC PROBE SPECIFIC CONSTRUCTION PROCEDURES

1. Mount the components on the perfboard as shown in Figure A-1b.
2. Solder the components together as indicated by the schematic diagram shown in Figure A-1a. Do not attach the alligator clips at this time.
3. Burn a small hole (for the probe tip) in the center of the solid end of the medicine bottle.
4. Burn two small holes (for the connecting leads) in the medicine bottle lid.

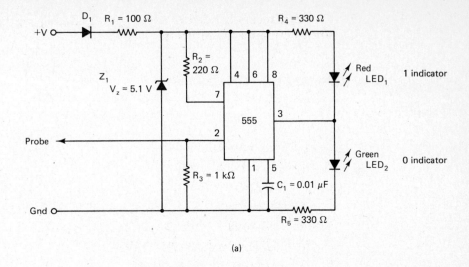

(a)

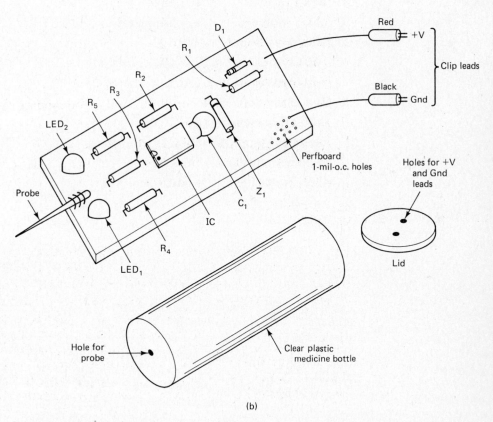

(b)

**Figure A-1**  TTL/CMOS logic probe: (a) schematic diagram; (b) construction layout.

5.  Carefully install the 555 timer IC into the IC socket.
6.  Place the circuit board, probe first, into the medicine bottle.
7.  Run the connecting leads through the holes in the lid and solder them to the alligator clips (take note that the $+V$ clip is red and the ground clip is black).
8.  Fit the lid on the medicine bottle, and the logic probe is ready for use.

## LOGIC PROBE OPERATION

Attach the red $(+V)$ clip lead to the positive voltage and the black (Gnd) clip lead to ground on the circuit to be tested. The green LED should be on when the probe is not placed on a circuit. Place the probe at the point in the circuit that you want to test for a logic condition. If the point is a logic 0, $LED_2$ (green) will stay on, and if the point is logic 1, $LED_1$ (red) will light.

## TTL/CMOS LOGIC PULSER

### LOGIC PULSER PARTS LIST

*1* 555 timer mini-DIP IC

*1* Eight-pin mini-DIP IC socket

*1* 1N4001 diode or equivalent $(D_1)$

*1* 1N5231 zener diode or equivalent with $V_z = 5.1$ V $(Z_1)$

*1* Red LED with $V_F = 2$ V $(LED_1)$

*1* SPDT momentary mini-pushbutton switch $(S_1)$

*1* 100-$\Omega$ resistor at 0.5 W $(R_3)$

*1* 220-$\Omega$ resistor at 0.25 W $(R_4)$

*1* 1-k$\Omega$ resistor at 0.25 W $(R_2)$

*1* 10-k$\Omega$ resistor at 0.25 W $(R_1)$

*1* 0.01-$\mu$F capacitor at 25 WV dc $(C_1)$

*1* Perfboard (approximately 1 $\times$ 3 inches)

*1* Aluminum probe or finishing nail

*1* Clear plastic medicine bottle that will accommodate the perfboard

*1* No. 22 gauge stranded-insulated wire (leads)

*1* Red mini-alligator clip

*1* Black mini-alligator clip

### LOGIC PULSER THEORY OF OPERATION

The TTL/CMOS logic pulser schematic diagram is shown in Figure A-2a and the parts layout is shown in Figure A-2b. Diode $D_1$ is used as a protection device in case the power supply leads are incorrectly attached to the circuit being tested. The logic pulser is used to inject a 1 or 0 condition into a digital circuit under test. The probe is normally at a logic 0 condition and the LED is off. When the pushbutton switch is pressed, the probe goes to a logic 1 condition and the LED with light. The 555 timer IC can safely operate over the voltage range +3 to + 18 V. When the pulser is being used on a TTL circuit with a power supply of +5 V, the 555 timer operates on a little less than 5 V and the LED will turn on normally. When the pulser is being used on a CMOS circuit with higher voltages, the 555 timer will operate normally, but the zener diode $(Z_1)$ operates and provides a safe +5 V for the LED.

### LOGIC PULSER SPECIFIC CONSTRUCTION PROCEDURES

1. Mount the components on the perfboard as shown in Figure A-2b.
2. Solder the components together as indicated by the schematic diagram in Figure A-2a. Do not attach the alligator clips at this time.

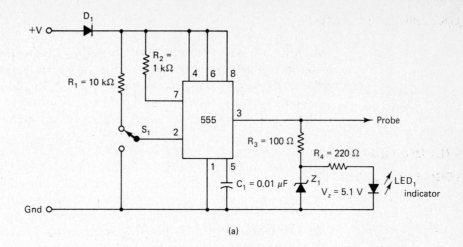

(a)

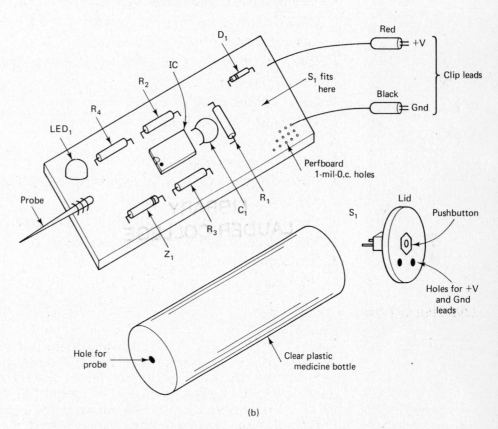

(b)

**Figure A-2**  TTL/CMOS logic pulser: (a) schematic diagram; (b) construction layout.

3.  Burn a small hole (for the probe) in the center of the solid end of the medicine bottle.

4.  Burn a small hole just above the center in the medicine bottle lid to mount the pushbutton switch. Mount the switch.

5.  Burn two small holes below the switch in the lid for the connecting leads.

6.  Using wire, solder the switch (with lid) as close as possible to the perfboard.

7.  Run the connecting leads through the holes in the lid and solder them

to the alligator clips (take note that the $+V$ clip is red and the ground clip is black).

8. Carefully install the 555 timer IC into the IC socket.
9. Place the circuit board, probe first, into the medicine bottle.
10. Fit the lid on the medicine bottle, and the logic pulser is ready to use.

## LOGIC PULSER OPERATION

Attach the red $(+V)$ clip lead to the positive voltage and the black (Gnd) clip lead to ground on the circuit to be tested. The LED should be off at this time. Place the probe at the point in the circuit where you want to inject a logic condition. If a logic 1 is desired here, press the pushbutton switch and the LED will light. If a logic 0 is desired here, first press the switch (LED on) and then release it and the LED will go off. The logic probe can be used at the output of the circuit to check for a proper indication.